ELECTROTECHNOLOGY VOLUME 7

ADVANCES IN ELECTRIC HEAT TREATMENT OF METALS

ELECTROTECHNOLOGY VOLUME 7

ADVANCES IN ELECTRIC HEAT TREATMENT OF METALS

NORMAN W. LORD
ROBERT P. OUELLETTE
METREK Division of the MITRE Corporation
McLean, Virginia
PAUL N. CHEREMISINOFF
New Jersey Institute of Technology
Newark, New Jersey

ANN ARBOR SCIENCE
PUBLISHERS INC / THE BUTTERWORTH GROUP

PREFACE

This book continues the series of studies initiated and sponsored by Electricité de France, surveying the status of electric technologies. The objectives have been to promote technology using electricity, and to diffuse information favorable to its development.

This volume reviews developments in electric heat treatment of metals to determine the outlook for market gain in some of the promising new technologies. The crisis in the fossil fuel supply creates higher costs for those fuels, and thus adds basic broad economic advantages to the use of electricity. As fossil fuel prices continue to rise, electric heat treatment technologies will become increasingly economical, despite their high capital costs.

Electrotechnology Vol. 2, Applications in Manufacturing (R. P. Ouellette, F. Ellerbusch and P. N. Cheremisinoff, Eds., 1978) presented the economic status and developmental state of electric technologies. Induction and other electric metal heating methods were discussed by G. Miller and M. Barbier. Several technologies, which at that time were regarded as only exotic laboratory tools, now have gained dramatically increased acceptance in commercial use, and are explored in greater detail in this volume.

The authors gratefully acknowledge Electricité de France for continuing its support in this work and sponsoring this study.

Norman W. Lord
Robert P. Ouellette
Paul N. Cheremisinoff

Norman W. Lord   **Robert P. Ouellette**   **Paul N. Cheremisinoff**

Norman W. Lord is a broad-spectrum physicist with the MITRE Corporation. Dr. Lord's research areas include the structure of solids, ocean acoustics and atmosphere-land interactions. He has consulted extensively for government and industries on a wide variety of problems in operations and estimating technological trends.

He received a BEE from Brooklyn Polytechnic Institute, and earned his MA and PhD in physics from Columbia University. Dr. Lord has been a research physicist at the Johns Hopkins University Applied Physics Laboratory, Hudson Laboratories of Columbia University and Travelers Research Center in Hartford, CT. He has written more than 40 technical papers covering these areas and his research, and is co-author of *Heat Pump Technology*, *Electrotechnology Vol. 4*, a 1980 Ann Arbor Science publication.

Robert P. Ouellette is Technical Director of the Environment Division of the MITRE Corporation. Dr. Ouellette has been associated with MITRE in varying capacities since 1969, and has been Associate Technical Director since 1974. Earlier, he was with TRW Systems, Hazelton Labs, Inc. and Massachusetts General Hospital. He was graduated from the University of Montreal and received his PhD from the University of Ottawa. A member of the American Statistical Association, Biometrics Society, Atomic Industrial Forum and the NSF Technical Advisory Panel on Hazardous Substances, Dr. Ouellette has published numerous technical papers and books on energy and the environment. He is a co-author of the *Electrotechnology* survey series published by Ann Arbor Science.

Paul N. Cheremisinoff is Associate Professor of Environmental Engineering at the New Jersey Institute of Technology. He is a consulting engineer and has been a consultant on environmental/energy/resources projects for the MITRE Corporation. A recognized authority on pollution control and alternative energy technologies, he is author/editor of many publications, including several Ann Arbor Science handbooks on pollution and energy, such as *Pollution Engineering Practice Handbook, Carbon Adsorption Handbook, Environmental Impact Data Book, Industrial and Hazardous Wastes Impoundment,* and *Environmental Assessment and Impact Statement Handbook*. He is a member of the Ann Arbor Science Publishers Editorial Advisory Board.

CONTENTS

LIST OF FIGURES

LIST OF TABLES

CHAPTER 1

INTRODUCTION

In earlier work,* the economic status and developmental state of electric technologies as practiced in U.S. industry were surveyed. Induction and other electric metal heating methods were discussed by Miller [1] and Barbier [2]. This Volume reviews the changes in technical development and economic status of all electric metal heat treatment methods that have taken place since the earlier work was completed.

For the electric heat treatment methods discussed in 1978, advances in available commercial technology essentially have been incremental. Electric heat treatment equipment has increased its market share primarily because of government-mandated reduction in the use of oil and improved economic circumstances for the use of electricity compared to natural gas. Induction heating and vacuum resistance heating, for example, have gained because their recognized technical advantages became much more cost-competitive as natural gas prices rose. With greater routine use a wider variety of special applications were tried and proven to further expand the market.

In part spurred by the increased familiarity with electric methods of metal heat treatment, several technologies, which were regarded in 1978 as only exotic laboratory tools, now have gained dramatically increased acceptance in commercial use. These technologies are laser and electron beam methods for localized heat treatment and plasmas for melting the more refractory metals. In addition, the economic status of electric arc melting for steel and ferro-alloys has become so much more favorable that conventional steel produc-tion, which relies on coal combustion, soon may be substantially displaced. Hence, these technologies have been added to the report coverage. The book therefore will cover the following major technical categories of electric metal heat treatment.

---

*Electrotechnology, Volume 2, Applications In Manufacturing.

- Electromagnetic Induction
- Lasers and Electron Beams
- Arc Melting and Plasmas
- Electric Resistance Heating
  - Vacuum and controlled atmosphere furnace
  - Resistance heating elements
  - Direct conduction heating

Notably absent from these is the use of infrared process heat. Most recently reviewed by Callaghan [3], it is apparent that its use is largely limited to nonmetallic materials, which are usually processed at much lower temperatures than are used in metal heat treatment. Infrared heaters may be fired by natural gas or may use electric resistance elements. In either case there is a low effectiveness of coupling to the metallic workpiece, which is not sufficiently compensated for by greater controllability.

The acceptance of electric methods by U.S. industry may be growing but there are still many influences that are unfavorable from a strictly individual company viewpoint. In addition, the motives of private industry leading to a positive interpretation of the financial balance among all the factors involved do not match those of the U.S. Department of Energy (DOE), which attempts to influence U.S. industrial energy use from a national energy use viewpoint.

For example, any new investments in the U.S. steel industry must compete with an existing plant capacity for steel-making which is demonstrably adequate for the market. Szekely [4] has discussed this barrier facing any new steel-making technology. The average U.S. cost of steel production is about $350/ton. Of this dollar figure, energy and raw materials comprise 45%, labor 35–45% and capital service charge 6–10%. Production of steel in the U.S. requires 36×10^9 J, 5 KWh/lb and 8.2 man-hours of labor. The book value of the plant is about $120/ton of annual output (less than one third the operating cost), but replacement cost is estimated between $1200 and $1500/ton. An automated arc furnace facility may reduce the energy and materials requirements, improve the product quality, particularly in ferroalloys, and drastically diminish required man-hours per ton produced. However, the capital service charge for this new plant would be $250–300/ton of annual output, completely overwhelming the savings in labor, increased product value and other benefits. Nevertheless, there are new plants being built because the oldest steel-making plants are much more expensive than the average, so that there is a realizable net return; also, individual company circumstances differ.

The situation in aluminum production as described by Brondyke [5] of ALCOA is somewhat different. Here, about 70% of the total energy consumed in aluminum production is used in the smelting process. It has always been feasible in aluminum production to introduce incremental energy saving improvements that have steadily reduced required energy. Brondyke cites the latest development of the ALCOA smelting process, which can produce aluminum at an expenditure of 4.5 KWh/lb (less than steel) compared to an ALCOA average of 7.5 KWh/lb. In the steel example, higher production, better quality and lower materials loss overcome a possibly greater energy expenditure for a net saving. In the case of aluminum, the saving is direct in energy and technically feasible without the high capital requirement of whole plant replacement.

The U.S. Department of Energy takes another viewpoint, which may be illustrated by comparing the ways in which it would view these two industrial initiatives. If no other factors are involved, a net energy saving is always regarded favorably. In ALCOA's case, DOE may be inclined to help support further technical development. If, on the other hand, there is a net increase in energy usage, then the Department considers how the change bears on its concern over domestic use of oil and natural gas. As Gross [6] pointed out, the use of oil is already under a strong absolute constraint for manufacturing processes. A conversion from natural gas to an alternate fuel (such as coal) may merit a tax reduction. On the other hand, in the case of the electrification of steel production there is, at best, no conversion since the utility may already use coal to generate supplied electricity and there is, at worst, unfavored fuel conversion if the utility increases its use of oil or gas to meet the new demand.

The foundry industry has been pointed out by O. Cleveland Laird [7] of the Department of Energy as one of the best examples illustrating the balance frequently struck between increased productivity and increased energy costs. Foundries generally are owned and operated by small companies. For the last 10 years they have been shifting fuel usage from oil and gas to electricity. However, James Williams [8], of Grede Foundries in Milwaukee, said that this trend has slowed recently due to a shortage of capital. The furnaces costs have been running $100,000-200,000/ton of melt capacity. This is a large investment for the small family-run company, which is still largely the rule in the foundry industry. At the same time, the cost rise for natural gas and propane, which had been outpacing that of electricity, has been slowed. At current price levels of 30¢/therm (100,000 Btu) for natural gas and 50¢/therm for propane, there is insufficient energy cost advantage offered by electric induction or resistance heating. The advantage would be in capability to automate the foundry operation. An example of this has been described recently by Layton [9]. Most of the product value of a foundry is represented by cost of energy used to melt and the capital cost for the furnace and pouring/molding equipment. Williams pointed out that since 1972 foundries have cut their specific energy usage 17-20%. For the small plants of less than 10 ton/hr, induction heating has emerged as a strong competitor of gas.

A very large company, facing many energy-requiring steps in its manufacturing processes, has a much more complicated decision to make, as pointed out by Gerhard Stein [10] of General Motors. This company utilizes a wide variety of energy-using equipment, and selection of any particular heating method must be made in the context of the company's best overall interest in its energy use profile.

For example, Table 1-1 shows the gross comparison for GM as a whole between 1972 and 1978 for electricity, gas and steam fuel. Production of vehicles at GM increased from 6.2 million in 1972 to 7.7 million in 1978. There was a reduction in energy requirements of 20% per vehicle because total energy usage was about the same—206 × 10^{12} Btu vs 203 × 10^{12} Btu. The proportion of electric energy usage increased from 24% to 29%. However, this change reflects primarily the large reduction, per vehicle, of gas and steam fuel usage compared to a small reduction in electricity usage, rather

Table 1-1. Energy Use in GM Components of Energy Consumption in GM [10]

	1972		1978	
	TBtu	Percent of Total	TBtu	Percent of Total
Electricity	49	24	59	29
Process Gas	70	34	62	31
Steam Fuel	87	42	82	40
TOTAL	206	100	203	100

Consumption per Unit of Production (Millions Btu/Vehicle)

	1972	1978	Percent Change
Electricity	7.9	7.6	−4
Gas	11.3	8.1	−28
Steam	14.0	10.7	−24
TOTAL	33.2	26.4	−20

Average GM Energy Cost

	1972		1978		Percent Change in Cost
	$/MMBtu	Ratio[a]	$/MMBtu	Ratio	
Electricity	3.50	5.5	8.50	4.1	143
Process Gas	0.64	1.0	2.09	1.0	227
Steam Fuel	0.62	0.97	1.86	0.89	200

[a]Relative to process gas.

than fuel substitution. Energy costs changed least for electricity in the 1972–1978 interval but, according to Williams [8] and others experienced in industrial fuel costs, the short-term outlook for natural gas is parity with electricity in rate of price rise.

The relative energy usage of General Motors in 1977 for its major processing categories is shown in Table 1-2. Metal heat treatment would be included only under the categories of metal casting, furnaces, ovens and metal forging. For these the total electric use is 12.8×10^{12} Btu, compared to gas use of 41.1×10^{12} Btu. Most of the substantial overall electric energy use is for space conditioning (HVAC), lighting, compressed air, metal machinery and assembly. It is apparent that electric methods of heat treatment are economical for a large firm only where there are special technical requirements that confer on them a high contribution to product value that offsets the higher cost of electric energy.

The distribution of energy use by temperature of the major processes, as shown in Table 1-3, shows that such special requirements are more likely to

Table 1-2. 1977 GM Energy Consumption by Form and Process
(trillions of Btu) [10]

Process Group	Steam	Gas[a]	Electricity	Total	Percent of Total
1. HVAC	39.6	12.3	10.0	61.9	31
2. Metal Casting[b]		11.1	5.8	16.9	8
3. Liquid Heating	19.4	1.2	1.3	21.9	11
4. Furnaces		16.3	2.2	18.5	9
5. Ovens	1.8	13.7	1.8	17.3	9
6. Paint Systems	3.2	6.0	2.6	11.8	6
7. Steam System Losses	10.6		0.6	11.2	5
8. Lighting			8.1	8.1	4
9. Compressed Air	1.2		6.1	7.3	3
10. Metal Machining			6.1	6.1	3
11. Assembly & Materials Handling		0.2	3.2	3.4	2
12. Plastics & Rubber	1.1	0.1	2.1	3.3	2
13. Metal Forging			3.0	3.0	1
14. Product Testing	0.1	1.9	0.7	2.7	1
15. Soldering & Welding		0.4	2.0	2.4	1
16. Forge & Forming	1.0	0.1	0.3	1.4	1
17. Domestic Hot Water	1.1	0.1	0.1	1.3	1
18. Other	2.4	0.3	1.7	4.4	2
TOTAL	81.5	63.7	57.7	202.9	100
Percent of TOTAL	40	31	29	100	

[a]4.3 TBtu of oil is included in gas figures.
[b]21.4 TBtu of coke is not included.

Table 1-3. Major Process Categories at GM by Temperature [10]

Process Group	Steam	Gas	Electricity
High Temperature			
Metal casting	–	9.7	5.8
Furnaces & heat treat	–	15.8	2.2
SUBTOTAL		25.5	8.0
Medium Temperature			
Ovens	1.8	13.6	1.8
Low Temperature			
HVAC	39.6	12.1	10.0
Liquid heating	19.4	1.2	1.3
Paint systems	3.2	6.0	2.6
SUBTOTAL	62.2	19.3	13.9
TOTAL	64.0	58.4	23.7

be found when high temperatures are required. Medium- and low-temperature applications usually involve heating air, which can be done more economically by steam or gas. It is when high temperatures must be induced directly in the material being processed (usually metal), and more so in a localized region of this material, that electric methods are most valuable. This is shown most clearly in Table 1-4, which breaks down the high-temperature energy consumption by specific processes. Melting, holding, die casting and hardening have the highest use of electricity compared to gas. Even here, however, there is some doubt about the economic force underlying electricity's share. Chandler [11] has cited a survey by the American Die Casting Institute, which attributes increased reliance on the electric furnace for holding (molten metal) to reserving restricted gas supplies for melting operations.

Stein expects that the outcome of competition between gas and electricity in metal treatment will depend on the economic performance for the specific company. This will include capital costs, operating costs other than fuel, and fuel. The relative significance of capital and nonfuel operating cost for electric and gas systems probably will remain the same over the short term except for a slight competitive edge by electricity due to its generally lower labor content. That edge is not expected to be decisive so that the competition will be decided mainly by fuel costs and fuel uncertainties.

Under the incremental pricing provision of the 1978 Natural Gas Policy Act, the U.S. government can arbitrarily adjust the industrial process gas price

Table 1-4. High-Temperature Process Energy Consumption at GM [10]

Process	Gas (TBtu)	Electricity (TBtu)	Total (TBtu)
Metal Casting			
Melting	8.9	4.4	13.3
Holding	0.5	0.6	1.1
Core making	0.7	0.1	0.8
Die casting	0.9	0.6	1.5
Briquetting	0.1	0.1	0.2
SUBTOTAL	11.1	5.8	16.9
Furnaces & Heat Treat			
Atmosphere generators	1.8	—	1.8
Kilns	0.1	0.2	0.3
Annealing	4.4	0.2	4.6
Hardening	4.6	1.3	5.9
Forging furnaces	5.4	0.5	5.4
SUBTOTAL	16.3	2.2	18.5
TOTAL	27.4	8.0	35.4
%	77	23	100

to be equivalent on heating value to that of Number 2 fuel oil, which current-ly sells in the U.S. for $5.00 per million Btu. If this policy were implemented during 1980, it would immediately make electricity costs less than two times the cost of natural gas. As indicated by some of the economic comparisons for induction heating in Chapter 3, this would tilt the cost balance decisively in favor of electricity for a large number of new applications. There may, of course, be a countervailing electricity cost escalation due to remaining utility use of natural gas. This may be mitigated and even reversed by new coal and nuclear-based power plants. Another major concern is the availability and reliability of natural gas supplies. Availability itself will influence heavily the government's decision on implementing the incremental price provisions of the Natural Gas Policy Act. At present there is an apparent gas surplus. How-ever, new additions to the known U.S. reserve continue to be less than annual production. Hence, shortages may well reoccur and, since the U.S. govern-ment does not regard industrial natural gas use as high priority, there will be even stronger curtailment of its availability for industry.

Electric heat treatment has its primary advantage in controllability, reduced labor costs in operations, localization of applied energy in some forms, and easier attainment of very high temperatures. On economic and environmental grounds, its chief competitor is the use of natural gas. However, the economy and availability of gas face an uncertain future. Hence, there is the widespread expectation in U.S. industrial circles of continued gains in the use of electric heating methods.

Chapter 2 presents a generalized economic analysis applicable to manufacturing energy systems. It will serve to establish a common terminology for describing the economic rationale underlying the new technical applications discussed in succeeding Chapters. Chapters 3-6 treat the major technical categories as organized above. In each section the major new technical developments since 1976 are described first. These are followed by smaller items, indicating promising new technical directives and some economic aspects of the outlook for the particular category of electric heat treatment. Finally, Chapter 7 describes the general market outlook for electric heat treatment.

CHAPTER 2

ANALYSIS OF ENERGY SUPPLY SYSTEM INVESTMENTS

As a practical matter, all investment decisions ultimately reduce to comparative evaluation of alternatives. If we use some single criterion as an aggregate of relatively weighted criteria, applied equivalently to all alternative options, we can compare all options to determine unambiguously that best option. Analysis in general terms may be carried out as a pairwise comparison of all alternatives. In trying to assess the relative susceptibility of different industries to new investments, based on their characteristic process energy requirements, our attention usually is centered on the effect of introducing a new energy supply system compared to retaining the prevailing practice.

The economic influence and acceptability of introducing a new electric heat treatment method can be treated by dividing its financial consequences into two parts. The first part includes those influences on all recurring costs and production capacity that vary with total production, plus influences on fixed costs that are independent of production. The second part includes all costs associated with the entailed capital investment. Comparison of these costs with anticipated benefits are calculated, in the first part essentially, to determine the acceptability of the investment and also indicate how anticipated future changes in costs and market may alter the acceptability.

COSTS AND PRODUCTION INFLUENCES

Usually there are three sets of manufacturing parameters changed by the new system that have the greatest financial impact:

1. expenditures on energy, material, operations and maintenance, as influenced by choice of energy supply;
2. unit production rates of the entire plant as influenced by required space, labor capability and utilization of production equipment; and
3. functional capability, quality or any intrinsic change in the nature of the product with a presumably positive market value.

Changes in the first two of these parameters are fully quantitative and tractable in a purely analytical approach. Often, although energy costs are a small part of the operation costs, the impacts of a change to electricity on other operational costs and the plant's production capability become the dominant factor in the comparison.

In the third parameter, a difference in functional capability also may be related to an appreciable difference in the product's market. Sometimes, however, the change cannot be expressed numerically. Quality and profound change in the nature of a product usually cannot be so expressed. It is then purely speculative to predict acceptable price levels and attainable unit sales volume. Nevertheless, in new applications this last aspect, with the promise of incalculable and, therefore, possibly very large, economic advantage, becomes the point that is most strongly expressed.

To be reasonably objective, the analysis in this section is confined to only the first two parameters. If favorable at this level, any uncertainty in the calculated costs must be weighted intuitively relative to any changes in the third parameter, which augur for increased business. In the ensuing section there are, in fact, few examples of electric heat treatment methods making available entirely new products. Hence, the analysis, based primarily on changes being sought in operating expenditures and production, will be very serviceable in comparing heat treatment methods when prevalent cost relations among the factors of production are used for each application.

We may derive the following cost elements and represent them symbolically for two circumstances: without any change and with the proposed change to an electric energy supply system. For element x the cost will change by the ratio $1 + a_x$:

Manufacturing Cost Element	Without	With
Total Unit Production Cost	U	$(1 + a_U)U$
Plant Unit Production Rate	P	$(1 + a_p)P$
Average Unit Energy Cost	eU	$(1 + a_e)eU$
Cost for Materials, Operation and Maintenance (O&M)	O	$(1 + a_O)O$
Fixed O&M Costs	F	$(1 + a_F)F$

In the abstract sense there are only two kinds of costs listed above: those that are associated only with each unit being produced and, therefore, vary directly with production rate, and those that are associated only with passage of time as the plant's physical existence is maintained, therefore independent of the level of production. The test of whether the proposed electric heat method is a plausible candidate lies in the value a_u, which we find by solution of the two equations expressing the balance between total unit cost and the sum of the cost elements. We tacitly assume that the increase in plant production can be sold. Otherwise, the problem becomes more global and we have to take into account a loss due to investment in idled plants.

Without the proposed energy supply system, the total unit production cost is related to the cost elements according to the equation

$$U = eU + O + F/P \tag{1}$$

On the other hand, with the proposed energy supply system,

$$(1 + a_u)U = (1 + a_e)eU + (1 + a_o)O + (1 + a_F)F/P(1 + a_p) \tag{2}$$

Subtracting Equation 1 from 2 yields the solution for a_u:

$$a_u U = a_e eU + a_o O + a_f F/P(1 + a_p) + \frac{F}{P}\left(\frac{1}{1 + a_p} - 1\right)$$

The total annual yield of the investment will then be

$$-P(1 + ap)a_u U = (A_p - a_F)F - P(1 + a_p)(a_e e + a_o O) \tag{3}$$

For a typical electric application, greater convenience, ease of material handling and reduced space requirements usually reduce unit operational and fixed costs so that a_O and a_F are negative and a_p is positive. However, energy costs are usually higher so that a_e is positive. Hence, the attraction of the investment lies in how far the productivity increase and operational convenience can overcome the negative influence of increased energy costs on the annual yield.

CAPITAL INVESTMENT CONSIDERATIONS

The relative attractiveness of any investment is based primarily on the ratio of the annual yield to the required capital cost. This will be modulated of course by anticipated future changes in the annual yield and the expected useful life of the equipment being built or purchased. This useful life may depend not only on the physical durability of the equipment but also on its promise of retaining its technical and economic edge in the face of future innovations or changes in marketability of the goods being produced. The most common U.S. industrial measure of such capital investment is the payback period, which is roughly the ratio of capital investment to annual yield. Usually, because of uncertainties about future technological development and markets, this period should be less than five years. In periods of generalized economic uncertainty, such as the extraordinary inflation of recent years and the current U.S. recession, requirements on new investment can be considerably more stringent. For example, as stated by Gonser [12], due to the sharp auto sales reduction General Motors is cancelling many development projects in heat treatment, retaining only those that promise to repay investment within one year.

A somewhat more formal criterion, usually offered by financial analysts, is the internal rate of return. Here one finds the rate of discount for the present value of future yields, which will cause their summation to just equal the required current capital investment, D. The present value, P, of future yield, $Y(t)$ ($-P(1 + a_p)a_u U$ in Equation 3), up to a specified time, T, continuously discounted at a rate, a, may be expressed as

$$P = \int_0^T Y(t)e^{-at}dt$$

The discount rate, a, is usually considered to be the interest rate equivalent to the combination of debt and equity financing characteristic of the organization considering the capital investment, D.

During the past years, U.S. prime interest rates have been as high as 20%. Long-term rates for corporate bonds have been well over 10% and, with common stock share values below book value, most firms are reluctant to finance new capital investment by issuing new shares. Hence we can expect (a) to be at least 10% so that contributions to present value, P, by $Y(t)$ are minimal for t greater than 10 years.

Conversely, if $Y(t)$ and D are known, one can solve the equation for that value of discount rate that would cause P to equal D and then compare it to the organization's financing cost. If one makes no speculation about future changes, such as inflation, and assumes the yield to be a constant, Y_0, the equation for P integrates to

$$P = \frac{Y_0}{a} (1 - e^{-aT})$$

For equipment lifetime, T (assumed above to be 20 years) and assumed constant return, Y_0, the internal rate of return is then the solution, a, to the equation

$$\frac{D}{Y_0 T} = \frac{1 - e^{-aT}}{aT}$$

Values of the internal rate of return, for representative values of Y_0/D when T is 20 and 10 years, are shown in Table 2-1. Marked changes in rate of return as yield elevates to the range of acceptable payback periods (4-6 years) essentially justify the superficially rough-and-ready investment criteria of

Table 2-1. Dependence of Rate of Return on Ratio of Yield to
Investment and Anticipated Useful System Lifetime

Yield/Investment	Lifetime (yr)	Rate of Return (%)
0.05	20	0
0.08	20	5.6
0.10	20	7.95
0.10	10	0
0.125	20	11.2
0.16	20	15.3
0.16	10	10.3
0.25	20	24.9
0.25	10	22.3

U.S. industrial firms. Generally, they seek to exceed by a fair margin a weighted debt interest charge representing their customary mix of corporate debt and stockholder equity. At the present time, this would be in the range of 8–9%, roughly approximating current rates on long-term U.S. Treasury bills. At a 10% yield, the rate of return for a 20-year system lifetime does not quite match that of a very safe bond. Cutting the system life to 10 years (due either to obsolescence or to even unanticipated wear characteristics in a new system) completely vitiates the investment value. On the other hand, for a 16% yield, we still have a robust return of 10.3% when the system lasts only 10 years. A 4-year payback is barely touched by the finite system lifetime.

In the above discussion, some ancillary considerations specific to U.S. circumstances have not been considered. For example, there may be available some rapid depreciation schedules for federal income tax accounting that will, in essence, reduce finance costs, the tax deferral acting as an interest-free loan from the government. The effect of this, however, is to reduce by only a few percentage points the requirement on rate of return that would decisively favor the new investment. In terms of payback period, the objective could be raised from 5 years to 8 years. A reasonable gauge for marketability of some of the new heat treatment methods and applications discussed in the subsections below would be as follows:

Payback Period	Acceptance
Greater than 8 years	No
Between 5 and 8 years	Maybe
Less than 5 years	Yes

CHAPTER 3

INDUCTION HEATING

The use of electromagnetically induced fields for heating metals was already well established from a technical standpoint in 1976. However, gas and oil were still so inexpensive that, unless the convenience and controllability of induction heating were absolutely essential, they were generally favored in customary economic evaluation. Since then there have been no major technical advances to support induction heat. The future of oil and gas use has been placed in considerable doubt. Hence, somewhat in advance of eventual truly favorable economics, the shift toward induction heating of metals has accelerated. The complex of reasons has been reviewed by Obrzut [13]. He cites a general change in attitude away from what used to be customary: that induction heating and melting of metals is an extravagance that is affordable only in the most exotic processes when its use confers a product quality advantage.

Many manufacturers have been compelled to use induction heating when their gas supplies were cut off. It then becomes the existing investment that is maintained in the face of a remaining slight and declining economic advantage for gas when it is available. It is also being chosen for very high production rate procedures. The applications where it offers a clear present economic benefit are growing. Some of the major forces in favor of induction have been changes in the automobile industry, such as the size reduction in cars, requiring new designs with surface hardening of smaller parts to compensate for reduction in bulk strength.

The major factors contributing most to induction heating growth are still its inherent process advantages, as outlined recently by Osborn [14], for example:

1. Increased Production
   Rapid localized heating reduces heating time.
   In surface hardening, need not require plating, packing, carburizing, straightening and other forming operations.

Compact equipment can be located close to related operations and materials supply points.

Process rearrangements for different parts are greatly facilitated.

Operator is "paced" by the equipment or the operation can be automated.

2. Improved Product Quality

Distortion is minimized in local heating and hardening; fewer rejects.

Scale is reduced due to shorter heating time; scrap is reduced by closer heating control.

Ductility of core is maintained in surface hardening.

There is accurate control of heated region for correct temperature and production uniformity.

3. Lower Costs

Labor hours are reduced; some skilled labor not required.

There are reduced floor space requirements with improved working conditions.

Maintenance costs are lower.

No warm-up time is required.

Plant operation is cleaner.

The induction heating equipment manufacturers have standardized power supply frequencies so that probably 99% of induced field power will be at the frequencies listed in Table 3-1, according to the type of equipment used to convert from the basic 60 H_Z supply frequency and the conversion efficiency from the field energy to the workpiece [14,15].

The vacuum tube oscillator transforms three-phase power line voltage to 15–20 kV, rectifies it and then, using vacuum tubes and reactance elements, including inductor and workpiece in the tank circuits, generates a high-frequency electromagnetic field whose frequency and intensity are strongly dependent on the dissipative load presented by the workpiece. The motor generator set usually consists of a conventional three-phase induction motor driving a high-frequency single-phase alternator. Frequency multipliers are the most efficient category and consist essentially of special transformers that triple the line frequency to single-phase 180 H_Z, which can be tripled again to 540 H_Z.

Frequency converters are the most recent entry and are gradually becoming the main source of high-frequency power when it is measured as total kilowatts installed each year. They use diodes and silicon-controlled rectifiers

Table 3-1. Standardized Categories of Induction Heating Equipment [14,15]

Type	Frequencies (kHz)	Percent Efficiency of Inversion	Power Ratings (kW)
Vacuum Tube Oscillators (high)	1000–4000	<50	<10
Vacuum Tube Oscillators (low)	200–450	50–60	5–600
Motor Generator Sets	1, 3, 10	75–80	7.5–500
Frequency Multipliers	0.180, 0.540	90–95	100–1000
Frequency Inverters	0.5, 1, 3, 10, 50	85–95	50–1500

(SCR) to first rectify primary 60 H_z and then shift to high frequency with SCR and reactance units. Some manufacturers convert directly from 60 H_z to as high as 3 kH_z. The shift to frequency inverters is basically economic because equipment cost in dollars per kilowatt is substantially less than that for the motor generator set. Above 10 kHz the cost rises so that it is appropriate only to special applications.

Operation and control of induction heating systems can be carried to high levels of flexibility and precision using modern electronic and electromechanical devices. Such devices can regulate power levels within 1% and application times to tenths of a second, if necessary. Frequencies can be adjusted to a small degree by variable capacitors or inductors in the tank circuit. Frequently there are several widely different frequencies available through switching arrangements. For many applications, essentially the same results can be achieved with more than one frequency by using different power densities and heating times.

The biggest and broadest impact of induction heating has been in the forging industry. Forging is the application with the clearest case for economic advantage over the predecessor heating method, natural gas, as described in the section entitled Induction Heating Economics. Forging is also an example illustrating the widest range of technical parameters at which induction heating is economically practical today. Frequency range runs from 50 Hz to 500 kHz, with power ranging from 60 kW to 2500 kW. George F. Bobart [16] of Westinghouse has estimated that new induction heating installations account for over 25% of total heating units purchased for forging furnaces.

Melting aluminum in electrical induction furnaces has been increasing due to its value in reclamation and the growing proportion of aluminum being used in automobile manufacturing. Induction for hardening always has been popular because of the selectivity in heater region and the advanced controls and automated processing that are facilitated thereby. One outstanding example, cited by Balzer [17] of Tacco, is a single-shot hardening machine for automotive transmission shafts. With a 6-second heat time, production rate is as high as 400 parts per hour.

The application of induction heating in steel mill processing began in the mid-1970s and, according to Joseph Chance of Ajax Magnethermics, appears likely to come to full fruition during the 1980s. Ways are being found to take advantage of the residual heat generated in the processing operations. In continuous casting, billets, bars or blooms are fed directly to the rolling mill rather than cooled after casting. Induction heaters restore the heat to the degree necessary for rolling; reheat energy requirements are cut by as much as 75%. In addition, the product items are more uniform due to closer temperature control. However, Bobart of Westinghouse cites only two known induction heating installations with continuous casters in the steel industry:

(1) at McLouth Steel, for heating slabs 30 ft × 5 ft wide and 12 in. thick; and (2) at U.S. Steel's Chicago plant for continuous cast 7.5-in. square bars in a direct inline casting-induction heating-rolling setup.

Probably the most difficult induction applications to develop are brazing and soldering, where both workpiece and filler metal are involved. This is, however, one of the most rewarding because once the working parameters in any specific application are set, a brazing or soldering process can then be incorporated as a precisely repeatable operation on a production line.

The next section outlines the main thrust of induction heating's economic advantage using the unequivocally favorable case of hot forging. Following sections describe the latest developments in the major application areas mentioned above. The last section discusses the current market outlook for induction heating technology indicated by overall industry statistics.

INDUCTION HEATING ECONOMICS

The economic disadvantages of induction heating are high capital equipment and electric energy costs. Equipment at several hundred dollars per kilowatt commonly cost several times that of fossil fuel-fired furnaces for the same applications [16]. Since 1976 there has been no reduction in these costs relative to those of fossil fuel except for oil. Hence, natural gas heating remains the principal fossil fuel competitor. Induction counters the fuel energy cost disadvantage by coupling the electric energy much more completely and precisely to the workpiece.

Heating by direct electric conduction also will couple supply energy tightly to the work, and the equipment is relatively inexpensive. However, it is appropriate only for through heating, requires electrical contact with the workpiece, causing often unacceptable problems in high-volume, continuous-process lines. Also, the workpiece usually must have a length to diameter ratio of at least 10 for good heating efficiency. The newer electric processes, such as electron beam and lasers, are very well suited for inline continuous production. However, their equipment costs are several thousand dollars per kilowatt and their energy conversion efficiency is very low. Hence, practical applications for electron beam and lasers are limited to unique process requirements. Induction heating, even as a mature technology with few new surprises to offer, remains the choice for the widest variety of metal heat treatments.

Hot forging steel is a major manufacturing process for which induction heating's economic advantage is well established and promises to grow greater with the anticipated outlook for fossil fuels. The underlying costs for the process are shown in Table 3-2. Prices for electricity and natural gas have risen by a little over 50% since 1977, as shown in part a) of the table.

Table 3-2. Underlying Costs in Hot Forging Steel

a) Industrial Energy Prices: U.S. National Average

Energy Type	1977 [18]			1980 [19]			
	Base Unit	$/10^6 Btu	$/10^6 kcal	Base Unit	$/10^6 Btu	$/10^6 kcal	Percent Increase
Electricity	$ 0.0216/kWh	6.34	25.15	$ 0.0329/kWh	9.66	38.40	52.3
Natural Gas	$ 1.60/mcf	1.52	6.04	$ 2.48/mcf	2.36	9.38	55.1
Residual Fuel Oil	$13.36/bbl	2.12	8.33	$24.50/bbl	3.90	15.52	84.0
Distillate Fuel Oil	$15.46/bbl	2.65	10.54	$31.30/bbl	5.38	21.40	103.0

b) Other Costs (10% annual rise from 1977–1980)

	1977	1980
Steel Product		
Value for scale loss	$350/ton ($366/metric ton)	$467/ton ($515/metric ton)
Value for scrap loss	$700/ton ($772/metric ton)	$934/ton ($1030/metric ton)
Production labor	$ 7/hr	$ 9.34/hr
Maintenance labor	$10/hr	$13.33/hr

Prices for oil nearly doubled. The other costs, shown in part 6), reflect simply an average annual inflation of about 10% over the three-year interval.

Those descriptive factors common to an exemplary plant hot forging steel are posed in Table 3-3. The annual output is assumed at 27,220 metric tons, with the plant operating somewhat less than half time at an instantaneous capacity of 6.8 metric ton/hr. To hot forge steel its temperature must be raised to 1230°C by supplying 189,900 kcal/metric ton. If fuel oil is used for heating it is assumed that distillate and residual would be used in the same proportion as used in the manufacturing industries as a whole. As seen in Table 3-2, both these fuels are more expensive in heating value than gas. The less expensive is residual fuel oil. However, there are extra costs included in its use due to pollutant control requirements and increased levels of metals and other corrosive or erosive impurities that damage the fuel pumps and pipelines. Hence, from a practical standpoint, the energy costs in using residual fuel are really equivalent to those in using distillate. At present, with roughly comparable capital equipment costs for distillate oil and natural gas use, only natural gas can compete with induction heating.

The distinctive factors of an induction heat hot forging plant are shown in Table 3-4. The range of frequencies that would be used depends on the thickness or diameter of the workpiece. Coupling effectiveness in forging would be between 50 and 65% [14,16,20] so that the net coupling from supply power would be, on average, 51.8%. The supply power required, which then would be 2900 kW, determines a 1977 capital cost at $580,000. The big

---

Table 3-3. Factors Common to Exemplary Plant Hot Forging Steel

- Instantaneous output   7.5 tons/hr (6.8 metric ton/hr)

- Plant usage factor   4000 hr/yr

- Annual output   30,000 tons (27,220 metric tons)

- Heat requirement to raise temperature of steel to 2,250°F (1,230°C)

  220 kWh/ton   or   683,000 Btu/ton
  (220.5 kWh/metric ton)   (189,700 kcal/metric ton)

- Industry reference usage of distillate and residual fuel in 1977

  Distillate - 103,000 bbl for 600 × 10⁹ Btu
  (151.2 × 10⁹ kcal)

  Residual 242,000 bbl for 1.520 × 10⁹ Btu
  (383.5 × 10⁹ kcal)

  Proportionate use for heat

  Distillate 0.282
  Residual   0.718

Table 3-4. Special Function of Induction-Hot Forging Plant

Frequency need	180 Hz–3 kHz
Efficiency in generation	85–95%
Effectiveness in coupling to workpiece	50–65%
Assume net coupling from supply frames at	0.90 × 0.575 = 0.518

Supply requirement: $\dfrac{200 \text{ kWh/ton}}{0.518}$ = 386 kWh/ton = 365,000 kcal/metric ton

Power = 386 kWh/ton × 7.5 ton/hr = 2900 kW

Capital Cost

			1977	1980
Per kW			150–250	
	avg		200	267
@2900 kW			580,000	774,000

Labor required	1 production worker full time 1/4 maintenance worker
Scale loss	0.5%
Scrap loss	0.25%

advantages of induction forging lie in the low labor requirements, one production worker and one-quarter maintenance worker, and low levels of material loss in the operation, 0.5% scale and 0.25% scrap.

Counterpart distinctive factors for a natural gas-heated hot forging plant are shown in Table 3-5. Average effectiveness of coupling the combustion energy is assumed as 17.5% [14,16]. Webley [20] considered the gas furnace efficiency range only 8–15% in the slot forge furnace, but 25–35% in the rotary gas-fired furnace. The difference, as shown below, could be decisive at current gas rates in minimizing the induction heating advantage so that it would not be favored either as new or as a replacement for an existing gas-fired unit. Energy supply requirement at the assumed efficiency of 17.5% is 7,380,000 kcal/hr. This results in a 1977 capital cost of $187,300 for the gas-heating equipment in the plant, using a rough average of $25,400 per million kcal/hr. Labor required is roughly doubled, and the material loss in scale and scrap is roughly quadrupled, according to Bobart [16] and Webley [20].

A comparison of hot forge operating costs for the two heating methods is provided in Table 3-6. Essentially, in 1977 and 1980 a small energy cost differential in favor of natural gas is offset by a larger differential in material loss in favor of induction. This is based on the set of values for scale and scrap costs by Bobart [16] for Table 3-2. However, Webley [20] does not

Table 3-5. Special Factors of Natural Gas Hot Forging Plant

Effectiveness of Coupling (gas combustion-heat to workpiece)

   15–20%

   Assume @17.5%

Energy Supply Requirement

   683,000/0.175 Btu/ton = 3,900,000 Btu/ton
                        (1,083,000 kcal/metric ton)

   @7.5 ton/hr require supply @29,000,000 Btu/hr
                             (7,380,000 kcal/hr)

Capital Cost

	1977	1980
Per million Btu/hr	$6,400	$8,540
Per kW	21.9	29.1
For plant	$187,300	$250,000

Labor Required

   2 production workers full time

   1/2 maintenance worker

Scale Loss

   2%

Scrap Loss

   1%

consider that scrap loss cost per ton is higher than that of scale loss. This would reduce the favorable induction differential by about $6/ton in 1980. Without the material loss differential, the small labor cost differential in favor of induction is not sufficient to offset the extra energy expense. In the case of natural gas, a third set of figures is added in the column labeled 1980 to reflect the potential influence of the U.S. government mandating that industrial gas error rise to the heating oil equivalent. With such a change, induction becomes substantially favorable, even in energy.

The evaluation of induction heating operational cost savings for investment consideration is outlined in Table 3-7. Webly [20] pointed out that with the gas-fired rotary furnace, in addition to high heating efficiency (30% plus 17.5% assumed in Table 3-5) the metal loss is reduced by one-third. These two changes would eliminate the induction heat advantage for 1977 and 1980 and reduce it considerably for the hypothetical circumstance 1980 (see footnote, Table 3-6). A roughly comparable situation would be a lack of any

Table 3-6. Hot Forge Operating Costs that Differ Between Induction and Natural Gas (per ton of product)

	1977	1980	Percent Increase since 1977
a) Induction			
Electric supply energy	$8.34	$12.70	52.3
Production labor	0.93	1.25	33.3
Maintenance labor	0.33	0.44	33.3
Scale loss	1.75	2.33	33.3
Scrap loss	1.75	2.33	33.3
TOTAL	$13.10	$19.05	45

	1977	1980	Percent Increase since 1977	1980[a]	Percent Increase since 1977
b) Natural Gas					
Energy	$5.93	$9.20	55	$16.84	184
Production labor	1.87	2.49	33	2.49	33
Maintenance labor	0.67	0.89	33	0.89	33
Scale loss	7.00	9.34	33	9.34	33
Scrap loss	7.00	9.34	33	9.34	33
TOTAL	$22.47	$31.26	39	$38.90	73

[a]Postulated natural gas costs if the price is federally mandated to rise to oil equivalent according to pro rata use by industry of distillate No. 2 and residual fuel oil No. 6.

Table 3-7. Annual Yield of Induction Heating Investment Compared to Investment (Y = yield; D = investment)

	1977	1980	1980[a]
1) Cost Difference Advantage of Induction per ton	$9.37	$12.21	$19.85
2) Adjusted Cost difference for No scale or scrap loss considered	−1.13	−1.81	5.83
Y_1 - Annual Yield for (1)	281,000	366,000	595,000
Y_2 - Annual Yield for (2)			174,600
Investment D_1 as difference between induction and natural gas	392,900	524,000	524,000
Investment D_2 as induction cost	580,000	774,000	774,000
Y_1/D_1	0.715	0.698	1.136
Y_1/D_2	0.494	0.473	0.769
Y_2/D_1	−	−	0.334
Y_2/D_2	−	−	0.226

[a]See footnote, Table 3-6.

metal loss differential shown in the table on row 2. To evaluate the cost savings yield of induction as an investment, it is compared both to investment differential between natural gas and induction and the whole induction investment. This latter case would correspond to the common situation of a preexisting gas-fired unit with substantial remaining lifetime. Ratios of yield to investment are, in all relevant cases, sufficient to encourage investment using the lower limit of approximately 0.25 discussed in Chapter 2.

The foregoing analysis can be reexpressed in terms of the generalized formulas discussed in Chapter 2 by considering the natural gas-fired unit as the existing equipment, with induction the candidate investment alternative. For 1980, the yields would be found from Table 3-6, using the right-hand side of Equation 3 in Chapter 2 directly as follows:

Since production and fixed plant costs remain the same, $a_p = 0$; $a_F = 0$, then

$$Yield = -P(a_e + a_oO)$$

$$a_e = \frac{12.70 - 9.20}{7.20} = 0.381$$

$$a_o = \frac{6.35 - 22.06}{22.06} = -0.713$$

Hence, the positive yield of the induction investment in these terms is due to a 71% drop in the nonenergy costs, which themselves amount to over 70% of the total production cost, at the expense of a rise of less than 40% in the energy costs.

The investment value of induction heating generally will turn on the question of how much it reduces nonenergy operational costs compared to the extra energy expense. For the slot furnace analyzed in the foregoing, the investment is favorable even at current 1980 prices. However, for the rotary hearth, gas-fired furnace, analyzed by Webley [20], the gas energy costs and nonenergy gas operating costs are both reduced so that the operating costs savings $a_o O$, are at best intrinsically greater than extra energy expense, $-a_e e$. In any balance situation the value of an investment requires a substantial margin, and this may not be clearly established for all operational situations. On the other hand, for 1980, the hypothetical case for raised natural gas prices, there is a savings even in energy, and the investment favorability becomes much more certain.

FORGING

The applications of induction heating in forging and similar processes span the whole variety of materials and operating temperatures. Table 3-8 reviews a sample of the parameters involved to illustrate the variety and compare the heat requirements for the different process categories. It is clear from parts a) and b) that hot forging imposes the highest energy requirements for each material. However, part c) shows the high electric energy requirements imposed by metals like copper with low electric resistivity and unit magnetic permeability. As described by Jennings [21], there are several ancillary aspects of induction heating that have not been figured into the preceding analysis of the section entitled Induction Hearing Economics. Facility of automation favors induction for high-volume forging operations. For identical operations the induction work coil does not change or require frequent re-design. On the other hand, in a small forging shop working on a variety of odd-shaped pieces, the size flexibility of a gas-fired furnace leaves it the better choice. In temperature control of a gas-fired furnace, optical pyrometers are used with temperature maintained within 12°C. With induction, much more precise regulation of power output and heating times is achieved easily. In practice it is often found that the force to shape a hot billet is less because of temperature uniformity and repeatability.

An automatic cycle induction forge machine that heats various sized billets to working temperature and positions heated stock into the first position of a forging press has recently been developed by Induction Process Equipment Corp., Madison Heights, Michigan [22]. The equipment was designed for the Federal Drop Forge Co. of Lansing, Michigan to handle billets 6.68 cm to 12.7 cm round corner square (RCS), with lengths ranging from 10.1 cm to 45.7 cm at production rates of about 5 metric ton/hr. Billets are manually loaded, fed end-to-end by a hydraulic push cylinder through six induction coils and an unload mechanism. Two separate power supplies are used to energize the induction coil line. The preheat supply is a frequency multiplier rated at 900 kW−180 Hz connected to the first two coils bringing the billet temperature up to the Curie point. The final heat up to the forging temperature (1260°C) is done by the last four coils connected to a Statipower solid-state inverter rated at 900 kW−1000 Hz. At the exit end the billet temperature is measured with a scanning optical pyrometer. If temperature requirements are met, the shuttle mechanism moves the billet laterally to a position in front of a final push cylinder, which then places it in the first forge position. If the billets are either under or over temperature, the shuttle mechanism moves them to a reject conveyor sorted according to their temperature state. The machine is shut down after a predetermined number of parts are rejected. Auxiliary electronic and thermal actuators control the billet position in the last coil to ensure proper heating and ejection from the coil.

Another automated forging operation has been described by Duff [23] of Mac Tools, Inc., in Washington Court House, Ohio. In this case, the automated forging is used to carry through the entire process of shaping professional hand tools such as screwdrivers and socket wrenches. Forming was

Table 3-8. Metal-Working

a) Temperatures Required for

Process	Steel	Required Stainless Steel	
		Magnetic	Nonmagnetic
Hot Forging	2250	2000	2100
Hardening/Aging	1700	1800	–
Annealing/Normalizing	1600	1500	1900
Warm Forging	1400	–	1200
Stress Relieving	1100	1100	1100
Tempering	600	600	600
Curing of Coatings	450	450	450

b) Physical Properties of Metals to be

Property	Carbon Steel	Stainless Steel	
		Magnetic	Nonmagnetic
Melting Point, °F	2700	2650	2500
Density, lb/in.^{3}	0.283	0.28	0.28
Specific heat (Btu/lb), °F	0.12–0.16	0.11	0.12
Resistivity, ohms ^{6}/cm^3	20–120	60–120	80–130
Thermal conductivity, Btu/(hr) (ft^2) (°F/f$_t$)	27–17	15	9–15

c) Average Energy Requirements for Induction

Process	Steel	Energy required, Stainless Steel	
		Magnetic	Nonmagnetic
Hot Forging	400	375	430
Hardening/Aging	250	260	–
Annealing/Normalizing	225	210	375
Warm Forming	175	–	250
Stress Relieving	150	150	200
Tempering	70	70	100
Curing of Coatings	50	50	75

[a]Based on inline continuous process.

Thermal Parameters [16]

Typical Metal-Working Processes

Processing Temperature, °F

Nickel	Titanium	Copper	Brass	Aluminum
2000	1750	1650	1500	1000
1400	1650	1500	1200	900
1700	1500	1000	1000	700
1200	–	–	–	–
1100	1100	500	550	700
600	600	–	–	–
450	450	450	450	450

Considered in Hot Metal-Working Processes

Nickel	Titanium	Copper	Brass	Aluminum
2500	3000	1980	1800	1200
0.3	0.162	0.32	0.31	0.098
0.11	0.125	0.10	0.10	0.250
10–45	60–120	2–10	6–20	3–11
20	9	225	80	120

Heating in Typical Metal-Working Processes[a]

kWh/ton, when material being heated is:

Nickel	Titanium	Copper	Brass	Aluminum
450	375	700	400	300
300	325	600	325	275
400	300	425	375	210
240	–	–	–	–
250	225	200	200	210
120	110	–	–	–
90	80	175	110	125

done on a Peltzer and Ehlers' Polymaster hot forming press, which has a rating of 175 tons, operating at speeds up to one piece per second on parts ranging up to 7.62 cm long and 4.5 cm diameter.

In considering the new press it was realized that induction heating would minimize labor and assure utilization of the maximum production rate of the press. Natural gas, at that time (1974-75), was still inexpensive and plentiful. However, its use would require too much heating time, produce too much scale, be dirty and noisy (from blowers and burners) and generate too much waste heat in a forge room already housing other slot-type forge furnaces. A Tacco induction heater, rated at 10 kHz and 300 kW, is used to heat cylindrical steel blanks to 1232°C as they pass through the induction coil. No changes are necessary to the controls on coils except when the blank size reaches 2.87 cm diameter. Changing to a larger coil takes 30 to 45 minutes. Temperatures are measured and controlled with an Ircon Automatic Optical Pyrometer.

The following desired operating parameters have been achieved for the combined facility of induction heater and forming press:

- High productivity—the average production rate is 1600 pieces per hour.
- Minimum labor—for example, one man can produce over 2 million sockets per year.
- Minimum scale on finished forgings—it is usually less than 0. 076 mm.
- Flexibility—most company products require only minor changes in tooling.
- Extended die life—in some cases die life has been multiplied 10 times, up to 50,000 pieces per die.

The Mac Tool Company anticipates that 90% of its socket press capacity will rely on induction heating.

To facilitate a rigorous billet-heating operation, Eaton Corporation is using zirconium oxide lines in its induction heating coils. The Glasgow, Kentucky plant of Eaton's Axle Division uses two 3-kHz induction heating systems to prepare steel-alloy billets to a precision forging press. The billets weigh 1.35 to 6.4 kg and measure up to 23 cm long and 7.6 cm diameter. One system heats billets to 765°C and the second to 1232°C. The zirconium oxide liners made by Corning protect water-cooled copper coils about one meter long. They last at least six months and have effectively prevented arcing between coil and billet, even when the first signs of deterioration became evident.

MELTING AND HOLDING

Induction heating is also proving economical for the melting and holding operations of foundries, chiefly for the way in which it facilitates automation of the foundry process. Notably, the introduction of induction heating has recently become favorable for iron in automatic pouring. Automation of foundry operation for the nonferrous metals such as aluminum has been a proven technology for many years.

The current status of automatic iron processing systems, using induction, has been reviewed most recently by Layton [9] of the Linberg Co. (now a unit of General Signal Co.). The requirement of significantly higher pouring temperatures compared to nonferrous metals has been the main factor retarding development and acceptance of such systems for iron. Some idea of the melting time and power required for melting steel can be gleaned from the melting performance curves of Radyne furnaces, shown in Figure 3-1. Induction melting for ferrous casting is done mainly for grey, ductile base and malleable iron. Added to automatic molding, sand preparation and reclamation, and casting cleaning, already proven technologies, automated pouring completes the development of the fully automated foundry.

Automatic iron pouring falls into two major categories:

- continuously moving mold lines
- indexing mold lines

In the first of these, a metal holding furnace is fitted with an automatically operated bottom stopper pour device that fills multiple-quadrant ladles with a measured amount of metal. These ladles, in turn, are mated automatically to a passing mold car containing a mold to be filled. After pouring, the quadrant ladle, mounted on a mechanized carousel, returns for refilling to the holding furnace. For indexing lines, the holding unit pours the molds directly if the mold line cycle rate, or indexing time, is greater than the fill time of the mold to be poured.

One system designed to pour consistent quantities of metal at uniform temperatures and specified pressure is the induction-heated Lindberg/Junker Automatic Iron Pouring System, illustrated in Figure 3-2. A refractory-lined furnace crucible is equipped with teapot filling and pouring spouts that provide for automatic pressure pouring of metal from the crucible. There is a single loop channel inductor heating the melt to replace the furnace heat losses and maintain constant metal temperature. Its capacity also includes some superheating capability. The pressure pouring system has solid-state computerized controls with a preprogrammed microprocessor to handle a wide range of requirements.

The furnace structure is mounted on a tiltable frame arranged for manual tilting to empty the furnace for shutdown, changing of alloys or for emergencies. There is a sensitive furnace weighing mechanism so that pouring may be controlled fully without any device in contact with molten metal. Reprogramming the microprocessor for various size molds is unlimited and can be done for molds that change in a fixed sequence or that may change in a random manner.

Some practical aspects in choosing among induction heating alternatives for

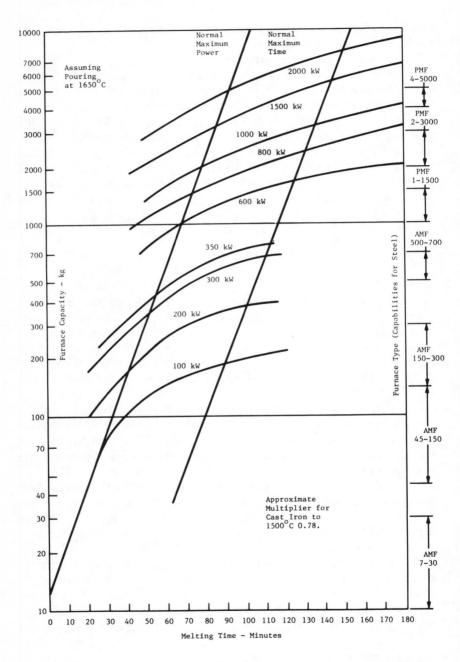

Figure 3-1. Typical performance of induction furnace for melting steel [13].

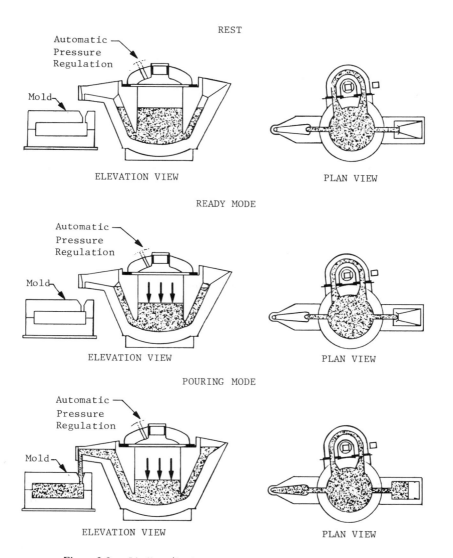

Figure 3-2. Lindberg/Junker automatic iron pouring system [24].

melting have been discussed by Schmidt [25] of the Tri-State Foundry Company in Cincinnati. The company had decided to use electrical heating and had to choose among electric arc, coreless induction and channel induction. The arc furnace was not justified because of its capital expense and the relatively heavy maintenance burden of the electrodes and furnace upkeep. The coreless induction melter was second to the channel because the melt batches were too continuous and Tri-State needed a lot of metallurgical control. The channel induction melter gave a much larger bath to work with. Also, the first cost of the coreless melter was much higher and it would sit idle on a third shift, which normally was used only to prepare a melt for the next day. A coreless furnace must be kept fairly full throughout the operation.

Tri-State chose a Lindberg/Junker-type channel furnace with an available pouring reservoir of 10.9 metric tons, a total melt bath of 13.2 metric tons and a melt rate of 1.13 metric ton/hr. The total enclosure of the hot melt helped minimize heat losses and controlled release of pollutants from the melt. Some of the advantages of the large bath are as follows:

1. A substantial flywheel effect makes it possible to tap out more metal than was charged in for a given period of time. Ideally, at the end of an 8-hour shift 10.9 metric tons would be tapped out plus metal charged at the rate of 1.13 metric ton/hr for a total of 20 metric tons. The furnace would continue to be charged on second shift and would be full by the following morning for first shift. This also meant lower demand costs since melting was spread out over more time.
2. There is better control of temperature and metallurgical properties. With a large batch they did not fluctuate from charge to charge. There was usually less than 0.45 metric ton variation among the batches.

There were also some specific disadvantages of the channel furnace:

1. It is difficult to remove slag. It must be raked or pulled out manually, and this can be done only when the furnace is full.
2. A somewhat limited mixing action or turbulence makes it necessary to use highly soluble graphite.
3. The furnaces operate continuously for approximately 6 months when they are shut down for inductor changes. This makes it necessary to have someone on duty 24 hours per day 7 days a week to monitor the furnaces.
4. There is a relatively long turn around time required for relining the upper case refractory, usually up to 2 weeks. This makes it advisable to have a standby unit to ensure continuous production.

A similar automated induction heated foundry system is marketed by Inductotherm of Rancocas, New Jersey, under the name AUTOPOUR. It is capable of pouring a mold every 10 seconds.

Some very large channel induction melting furnaces are available from Ajax Magnethermic. The Mueller Co. in Albertville, Alabama utilized a number of 50-ton, 2500 kW Ajax Magnethermic Jet Flow channel furnaces to spread electric power demand over a 24-hour period, rather than concentrating all melting during the day when charges are highest [26]. The Bridgman Casting Center of Bridgman, Michigan reaped a number of significant benefits with a 540-kW Ajax Magnethermic Vertiflex channel furnace to hold cupola iron [27]. It works with a 20-ton batch. The furnace can be filled and tapped simultaneously for uninterrupted production. The metal is tapped into 272-kg ladles to supply two rotary green sand lines and a single loop shell casting line. Prior to Bridgman's induction installation the company had problems with cold iron, poor chemistry control, high scrap losses and excessive downtime. These problems were essentially solved. The furnace, according to Paul Kays of Bridgman, is expected to pay for itself in two years. In a third AJAX example, the CWC casting division of Textron saw fit to replace a 6-ton electric arc furnace with a 50-ton Ajax Magnethermic Vertiflex hydraulic tilting furnace to hold, superheat and duplex grey iron from their cupola [26]. This experience confirms the judgment of Tri-State in Cincinnati, explained above, to select induction over arc for its automatic foundry. A small loss in thermal efficiency was offset profitably by elimination of electrode costs, 90% reduction in oxygen use by the cupola, 35% reduction in labor costs. 5.5% reduction in coke usage and a 72% reduction in downtime attributable to the melt shop.

The use of induction heating, compared to other forms of electric heating for aluminum casting, has been discussed recently by McKenna [28] of the Lindberg Co. (now a unit of General Signal Corp.). Prior to 1970 only 30% of Lindberg's furnaces sold for aluminum casting were electrically powered. In 1970 the proportion jumped to 50%, and the average over the last 4 years has been 75%. He finds that the immersion tube resistance heating element is induction heating's closest competitor. They are most ideal for aluminum and zinc die casting holding furnaces. In thermal efficiency, immersion heaters at 70% have a small edge over induction heating at 60%. This is for continuous operation, which always brings out the highest thermal efficiency. In melt loss, induction is slightly better than immersion. Within induction, McKenna finds some preference for channel induction, which has a much lower initial cost than coreless induction. In fact, for melting applications at about 454 kg/hr, channel induction can be almost as low as gas-fired in capital cost. However, to melt down the solid scrap on billets one has to use coreless. In the 1360 kg/hr melting class, channel or coreless induction are the only electric units applicable and, at that level, they are both more than twice the cost of comparable gas-fired equipment.

HARDENING AND TEMPERING

Induction heating for hardening and tempering steel is at once probably the oldest, most sophisticated and most unequivocally accepted application. The technical requirements are usually so exacting and the cost of required energy so low compared to the added value to the product being processed that the greater control afforded by induction easily makes it the choice over the closest competing heat source, natural gas. Hardening steel is accomplished by first heating a thin surface layer about 720°C to the austenite phase, as shown in Figure 3-3, to move extra carbon into the solution in this small

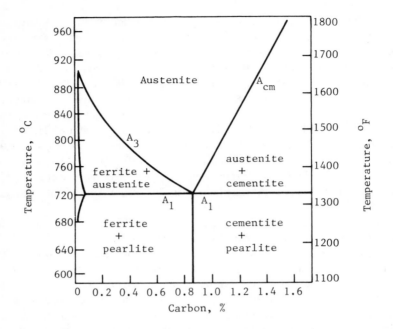

Figure 3-3. Critical temperatures in plain carbon steels and constituents, which are present in iron-carbon alloys on very slow temperature changes [29].

region. It is then quenched to freeze in the martensitic structure that is extremely hard due to the carbon-enriched cementite retained as a fine dispersion. The brittleness that accompanies hardening can be relieved in a controlled manner by reheating for a brief period at temperatures between 200°C and 649°C.

As described by Balzer [17] and Osborn [14], induction heating for surface hardening of large diesel crankshaft bearings was carried out in the early 1930s by the Ohio Crankshaft Co. using motor generators at 1920 and 3000 Hz. The journal was heated in 3.5 seconds using 2.32 kW/cm² of induced electric power for a hardened case depth of 4.5 mm. This depth greatly exceeds what is required to support the load or provide wear resistance. As equipment evolved over the years, a much wider range of available working frequencies made it possible to govern precisely the case depth and other important parameters of the hardened region. In Figure 3-4, the arrangement of inductor and quenching technique are illustrated for the simplest case of hardening the surface of a shaft or axle.

One of the most recent and difficult heat treating problems has been the result of requirements for the use of nonleaded fuel in automobile engines.

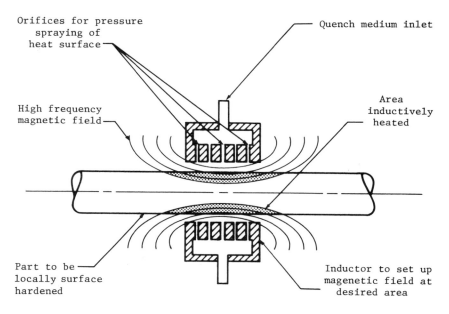

Figure 3-4. Hardening by induction heat and quench.

Within the combustion chamber, leaded gasoline forms lead oxide which, deposited on the valve seat, acts as a lubricant between valve face and valve seat area. With lead-free gasoline this lubricant is lost, and rapid wear, resulting in valve seat recession, occurs on the valve seat area. Pfaffman [30] of Park-Ohio describes the developments, by Tacco (subsidiary of Park-Ohio), of an induction heating technique to harden accurately to uniform depth the circumference of the seat area, regardless of the casting geometry variations. The process accommodates all normal variations in the cast iron material of the cylinder head and leaves no further machining to be required. Improvement in performance of the valves is illustrated in Figure 3-5. A specially designed floating induction coil was developed so that all valve seats in a cylinder head could be processed simultaneously.

IMPROVED CONTROL OF INDUCTION HARDENING PROCESS

Two recent developments show how induction heating is being utilized to carry out several hardening processes simultaneously. A numerically controlled dual heat station induction hardening machine that automatically

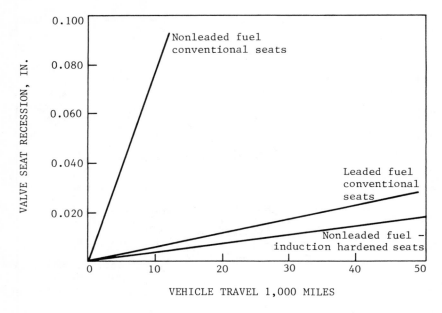

Figure 3-5. Valve seat durability with induction hardening and nonleaded fuel [30].

heats and quenches both the i.d. and o.d. of outer race bells of front-wheel-drive universal joints to an exact required hardness pattern has been developed by the Induction Process Equipment Corp., Madison Heights, Michigan. At the start of a hardening cycle the operator loads two outer race bells, one of each of two heat stations of the machine. In one heat station, the bell spline shaft is in a "down" position and in the other it is "up." The parts are automatically indexed up into their respective inductor coils and rotated by the numerical control system. The parts are rotated during heating to assure a uniform hardness pattern around the circumference. First one bell has its i.d. heated. After this operation power is switched to the station heating the o.d. of the other bell. During this heating the first bell is quenched. Next, the second bell is quenched to complete one half cycle. The bells are next lowered to be exchanged between heat stations. An identical second half cycle completes hardening both i.d. and o.d. of both race bells. In a second development a dual spindle induction scanner enables the progressive heating of two shafts simultaneously, either in different zones or end-to-end, as required. The shafts are hardened to a minimum core depth of 0.79 mm. The scanner is fully programmable to harden and spray quench various areas on the shaft in sequence without removing them. Changeover time for different shaft diameters is 20 minutes. The generator is a 50 kW unit of Cycle-Dyne Inc., subsidiary of Axel Electronics, Inc., in Jamaica, New York.

HARDENING GEARS

There have been some noteworthy recent developments in the heat treatment of gears. By contrast to the conventional practice, described by Arnold [31], requiring a separate quenching stage, Wolf [32] of Cincinnati Steel Treating Co. has described a system that allows induction hardening in the continual presence of a quenchant fluid. This is particularly advantageous for large gears where progressive, tooth-by-tooth hardening methods usually are employed. The gear is submerged in a temperature-controlled bath of quenchant. A specially designed induction coil with field intensifier is passed at the desired scanning rate through each gear tooth space, progressively heating a narrow section of the two adjacent flanks and the root. The case pattern and effective depth depend on the gear material, its prior treatment, the intensifier design and profile, induction frequency and power, scanning rate and the temperature, concentration and flow quenchant over the gear teeth. One example of thorough hardening of the nut and billet along with tooth face is a high-speed marine drive gear, 223 teeth, 8-diametral pitch, 13° helix angle, 5.1 in. wide and finished weight of 330 lb. It is produced from a 1200-lb AISI 4140 aircraft quality, electric furnace, vacuum degassed steel forging. The scanning rate is 14.5 in./min and the induction frequency is 150 kHz. The quenchant is a 6% solution of UCON A at 37.8°C.

A new dial-input numerically controlled induction gear hardening machine has been developed by Induction Process Equipment (IPE) Corp. of Madison Heights, Michigan [33]. It scans, heats and quenches tooth spaces individually to the exact required depth and hardness pattern without causing over-tolerance gear distortion. The inductor scans at a controlled traverse rate of about 6.3 mm/sec and induces heat only to that area of the tooth and root area being scanned. With this applied heat and a spray quench the gear is kept at the optimum temperature to minimize distortion and resulting grinding, finishing and cleaning operations. The power supply for the system is a new dual-frequency 75-kW, 25-kHz IPE Statipower, also operable at 9.6 kHz. The same company had previously developed an automated induction heating machine that anneals various sizes of welded starter motor ring gear blanks at a production rate of 1200 parts per hour [34]. The conveyor and loading equipment includes easy adjustment so that they can accommodate ring gear blanks ranging in size from 28 to 41 cm diameter. The same inductor coil is used for all annealing operations, to eliminate coil changing. Closed-loop deionized water recirculating, hydraulic power and automatic lubrication are fully integrated into the system. The inductor coil is energized by a 600 kW, 1000 Hz solid state Statipower inverter.

SHAFTS, CLUTCHES, CAMS AND OTHER COMPONENTS

The American Induction Heating Corporation of Detroit [35] has produced a custom designed facility that hardens camshafts automatically under completely controlled conditions. Inductor power is 300 kW at 180 Hz. Loading, positioning, heating and quenching cycles are controlled by the system with little attention. Alignment of the camshaft on carriage-mounted centers, onto which it has been loaded, is checked and adjusted, after which it enters the heating inductor. On reaching a sensed hardening temperature the camshaft X is returned to the load-unload position from which it is transferred to an agitated quench bath after a timed interval. When it reaches the preset lower temperature, the camshaft is then conveyed to a washer. Throughout, the cycle temperature is monitored so that it may be stopped if temperature ranges beyond preset limits. Accurate electromagnetic (em) field alignment is provided with a movable three-axis transformer base.

A new style of snapping head blade being treated is 47 cm long by 9 cm wide and made of AISI 1045 steel. It is a component of the corn head on an Allis-Chalmers "Gleaner" combine used to snap the ear of corn off the stalk. An induction heating unit at the Agricultural Equipment Implement Division plant of Allis-Chalmers Corporation in La Porte, Indiana is providing production savings and efficiency in the hardening of these blades [35]. The unit was designed and built by Cycle Dyne, Inc., as a single station-type 0-500, 50-kW generator arranged for use with two remote heating stations, set up adjacent to the forming press. The press operator fills the 400 blade-capacity hopper every two or three hours, and the induction heating operation proceeds fully automatically. Twin induction coils heat 5 mm of each side of the blade as it is fed down a channel. Hardness is monitored so that it is in the range Rc 48 to Rc 55. After hardening, the blades are dropped onto a conveyor for transfer to the forming press. Since only a 5-mm region on each side of the blade is hardened, the rest remains soft to facilitate the forming process.

New facilities for retooling hardening machines have been developed by the Industrial Process Equipment Corp. It uses one machine to perform a wide variety of hardening operations. A typical example is a numerically controlled unit designed and built for the Funk Manufacturing Company, Division of Cooper Industries, Coffeyville, Kansas. It hardens both shafts and sprockets: 2-cm–5.3-cm-diameter shafts up to 1.32 meters in length, and tooth and root areas of sprockets from 15.8 to 79 cm in diameter. A standard vertical induction hardening shaft scanner was redesigned with additional tooling, power supply and induction coil for sprocket hardening. There are adjustable tooling centers for shafts, a lift and lower rotating fixture for sprockets, a single-turn inductor coil for hardening shafts and a multiturn inductor coil for hardening sprockets. A special heat station includes an auto transformer outlet for sprocket inductor and an isolation transformer outlet for the shaft inductor. There is also a fully integrated recirculating spray quench for shafts, submerged agitated quench for sprockets, and a remote quench reservoir for storage of different concentrations of quenching fluids. Two frequencies of induced fields are available: 3 kHz for sprockets and 3–10 kHz for shafts.

The heat treatment of chain saw and small engine components was described recently by A. J. Craig [36] of Textron's Homelite Division in Charlotte, North Carolina. Generally deep, hardened regions are required, and induction frequencies of 400 Hz are used. Since natural gas is still in good supply in this area, the case for induction as a new investment is still not unequivocal. The best application for induction heating is in shaft bearings, where a relatively thin case depth is desired and energy cost is still not an appreciable part of total cost compared to labor. Induction provides production advantages in this area plus a limit on distortion that would occur otherwise and necessitate extensive regrinding.

STEEL-MAKING USE OF INDUCTION FURNACES

An apparent movement toward the use of induction in basic steel-making processes is indicated by recent plant installations, as well as the development of a new technique in oxygen-based steel-making.

INDUCTION AS REPLACEMENT FOR ARC FURNACE

One of the most effective furnaces used to melt or hold in the iron- and steel-working is the high-power arc furnace. By taking electric energy directly from the mains and introducing it only into the furnace interior, the inherently lower thermal efficiency of electricity compared to fuel combustion is countervailed to a large degree. At the same time the interior furnace chemical environment is much more easily controlled and the entire steel-making process can be rearranged for automatic operation. The obvious further alternative of induction heating is usually rejected because of an added loss of thermal efficiency in a highly energy-consuming process. Nevertheless, the CWC casting division of Textron saw fit to replace a 6-ton electric arc furnace with a 50-ton Ajax Magnethermic Vertiflex hydraulic tilting furnace to hold, superheat and duplex grey iron from its cupola [26]. This experience essentially confirms the parallel judgment of Tri-State in Cincinnati, explained above in a previous section, to select induction over arc for an automated foundry. The loss in thermal efficiency proved relatively small and was more than offset by elimination of electrode costs with attendant operational problems, a 90% reduction in operation use by the cupola, a 36% reduction in labor costs, a 5.5% reduction in coke usage and a 72% reduction in downtime attributable to the melt shop. In addition, the process is even better controlled and more easily made automatic.

INDUCTION AS REPLACEMENT FOR BLAST FURNACE

A while ago, it was disclosed that three 70-ton-capacity coreless induction melting furnaces would melt scrap steel and supply molten metal feedstock to a basic oxygen furnace [37]. The furnaces were made by Inductotherm Corporation of Rancocas, New Jersey, to be installed at the Natrona, Pennsylvania Melt Plant of Brackenridge Works, Allegheny Ludlum Steel Corporation. They have since been placed in operation with no more than usual startup difficulties over the fall of 1979 to winter 1980 period. At the present time they are in essence replacing a hot blast cupola for melting the charge of ore and scrap to feed the Basic Oxygen Furnace (BOF) shop. They melt at a rate of 75 ton/hr each, with 21.5 MW of power used in each furnace. The processing change eliminates foundry coke as an energy source (current cost $130–150 per ton) and provides inherent 80% efficiency of energy introduction with induction melting. The cupola was only 30% energy efficient and restricted in regard to size, shape and form of scrap that can be melted. It had also become very expensive to equip for meeting air and water pollution control requirements. The entire installation includes horizontal twin vibrating 80-ton/hr scrap preheaters made by Venetta of Warren, Ohio. These raise the temperature of cold scrap to 426°C to remove water and residual oils. Each of the three induction furnaces illustrated in Figure 3-6 uses a 6-cycle single-phase ac supply, so that there are no balancing problems as there would be in an arc melting furnace.

SUPERHEATING HOT METAL WITH AUXILIARY CHANNEL INDUCTION FURNACES

Superheating increases oxygen steel production and lowers its cost. The process has been developed by the Energy and Materials Corporation and uses the channel induction-type furnaces developed by the Ajax Magnethermic Corporation [38]. The process calls for the substantial superheating

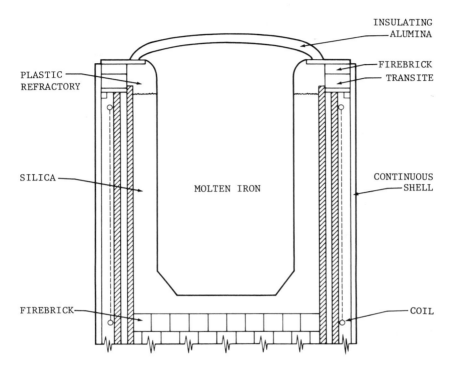

Figure 3-6. Induction furnace for heating charge to BOF.

of blast furnace hot metal within the time cycle associated with the production of a basic oxygen melt and charging this superheated metal into the vessels to melt more scrap and other metallics. By raising the temperature 500°F in a channel-type induction furnace, average capacity of a BOF and Q-BOP vessels can be increased by 13–15%; while at the same time reducing operating costs for materials at all levels. Scrap, pellets, sinter, roll scale or cold iron can be charged, depending on availability and relative costs. Installation of the induction furnace should allow charging directly from the blast furnace hot metal transfer ladle. The superheated metal should be discharged directly into the basic oxygen transfer ladle. There are also some advantages in practice, which can result in substantial but less predictable savings, such as the following:

1. Improve scrap allocation options.
2. Accurately control hot metal temperature, thereby obtaining better control of turndown temperature, flux and spar consumption, and improved refining practices, lining life and refractor costs.

3. Minimize the need for fuel-oxygen lances or costly additives, such as silicon carbide or calcium carbide aimed at melting additional scrap or other coolants and, also, related maintenance.
4. Minimize disruption in steel production arising from irregularities in iron production schedules.
5. Uneconomical slack wind blowing of blast furnaces can be eliminated by putting excess iron in inventory.
6. Hot metal storage and drawdown capacity of the "superheater" can be used to average out variations in iron analysis and to minimize schedule distortions due to differences in the range of blast furnace production and steel plant consumption of hot metal.
7. Low cost iron or steel scrap can be melted in the "superheater" with unused heating capability at incremental power rates, and these metallics also can be charged as supplemental coolants, as can pellets, coarse ore, sinter or other agglomerates.

"Superheating" by induction adds large amounts of energy to the BOF system efficiently. The "superheater" is a refractory lined, horizontal cylinder, similar in appearance and function to the conventional hot metal mixer. In addition, a series of six high-capacity, Jet-Flow channel-type induction heaters are attached to the underside of the furnace. Each induction heater operates on a single-phase, line frequency current and can provide a maximum of 2500 kW. Since only 28 kWh are required to raise the temperature of 1 metric ton of hot metal by 100°C (including all electrical and radiation losses), it then follows that a furnace equipped with six inductors (15,000 kW) could superheat hot metal at the following rates:

Hot Metal (metric ton/hr)	Superheat (°C)
536	100
268	200
179	300

With the high-capacity, Jet-Flow inductor, the 28 kWh/ton/100°C corresponds to an 84% conversion of kWh to kcal contained in the hot metal. If hot metal is "superheated" by 225°C, requiring 63 kWh/ton, the kilocalories imparted to the metal are sufficient to melt 137.9 kg of steel scrap and bring it to 1600°C, the approximate tapping temperature for the basic oxygen furnace. Table 3-9 shows the quantities of fuels required to also melt 137.9 kg of additional steel scrap, and the relative cost of each fuel (1977 U.S. $) compared with the 63 kWh required for "superheating" the hot metal. The "relative cost" comparison only takes into account the quantities and prices of the fuel and oxygen required, and the yield of raw steel from the 137.9 kg of steel scrap. Costs for additional fluxes (burnt lime and dolomite), increased furnace time and damage to the refractory lining are not included. The low yield of raw steel by the techniques using any form of silicon as a fuel results from the need to increase the flux addition (lime and dolomite) by 9 kg for every 1.0 kg of silicon to maintain the proper basicity of the slag

Table 3-9. Relative Cost of Alternative Sources of Energy for Melting
137.9 kg of Additional Scrap (USA-1977) [38]

Energy Source	Quantity	nm^3 Oxygen	kg Scrap Melted	kg of Steel	Relative Cost of Fuel
Superheating	63 kWh	None	137.9	131.0	1.0
Oil-Oxygen	13 liter	29.3	137.9	131.0	1.6
High-Silicon Iron	+1.09%	12.8	137.9	98.7	3.1
Calcium Carbide	19.5 kg	9.0	137.9	131.0	6.0
50% Ferrosilicon	19.4 kg	14.6	137.9	111.5	6.2
Aluminum	8.7 kg	7.7	137.9	128.4	8.4
Silicon Carbide	18.6 kg	24.7	137.9	93.8	9.0

in the basic oxygen furnace. Because of the resulting increase in slag volume (16 kg for each kg of silicon), there is an increased loss of 2.9 kg of iron to the slag for every extra kg of silicon added to the system. There will also be reduced labor costs per net total metric ton of steel production and an increase in the life of the furnace refractories because of the better control of the turndown temperature. The total of the decrease in cost for fluxes, oxygen, labor and refractories is usually slightly greater than the cost for the electric power required for inductive heating, leaving the difference in metallic cost as the net overall savings to be balanced against the induction heater capital investment.

BRAZING, SOLDERING AND SINTERING

The use of induction heat for brazing, soldering and sintering shares with hardening the feature that costs for energy involved in the process are a small portion of the added product value. Exactly what the economic advantage will be over the use of other energy forms depends markedly on the specific application. A recent review of brazing by Smith [39], for example, has discussed some of the particular requirements in design, processing specifications and proper selection of equipment. The following advantages of induction for brazing and soldering can be cited:

- precise control of heat in a local area,
- opportunity for semi or full automation of production process,
- observability of results by operator at all times,
- allowance of either preformed or hand-fed alloys,
- lowered contamination of workpiece from heat source, and
- lower consumption of energy.

However, these advantages, to be realized, require the careful coordination of joint design, brazing alloy and inductors, as in the examples shown in Figure 3-7. In part a), the important features are that the alloy is drawn up into the joint over the entire circumference by capillary action and that any excess can drain through the hole drilled in the outer lower member. Also, the inductive field heats the workpiece member being joined rather than the alloy. In part b), the bellows is allowed to retain flexibility by not having

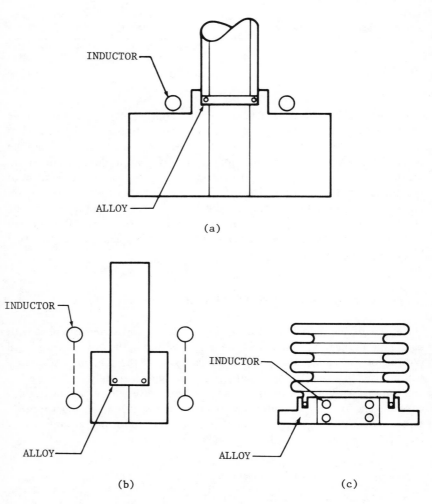

Figure 3-7. Recommended joining procedures in using induction heating.

excess brazing alloy accumulating in the flexonal region. The inductor is placed to heat primarily the massive flange member. In part c), the joint is designed with an unobtrusive sleeve that reduces the mass that must be heated by the inductor.

In another recent report, Larson [40] has discussed some special advantages of using 50 kHz induction fields. This frequency lies in the relatively under-utilized range of 10–200 kHz. It offers deeper penetration than 200 kHz without grain growth or retained austenite on hardened sections. Flux concentrators can be fashioned without using iron laminations. Powdered iron aggregates of 80% Fe, 20% plastic can be machined to fit around the inductor coil. The frequency is high enough to allow self-quenching in hardening processes. In this way it recoups one of the advantages of lasers and electron beams. Larson has found the frequency excellent for soldering, brazing, forging and heat setting of glues with epoxy casing. It has also proved a useful alternative in hardening exhaust valve seats, using a pulse of 4-5 seconds compared to one of 1-2 seconds. Much lower frequencies are found useful in fusing powder metal contacts, since much deeper penetration is required.

Robert L. Conta [41] has developed a practical technique to overcome the "incubation problem" in 3 kHz induction heating of sintered steel compacts. The material is preheated within the depth to be treated up to 400°F, using either radiant or radio frequency (RF) induction heating. This technique establishes the metallic continuity for eddy current circulation, which is normally lacking in the initial stages of the 3 kHz field application.

MARKET OUTLOOK FOR INDUCTION HEATING

Induction heating for a variety of electric heat treating processes has established for itself a substantial segment of the market for industrial heating equipment. There are obvious reasons, such as the uncertain supply and increasing price of fossil fuels, as well as greater productivity, less maintenance and a safer, cleaner working environment. During the 1970s induction heating broadened its competition with fossil fuels from mainly technical grounds to occupational safety and the environment. The last two areas acquired new force with the establishment of federal regulatory authority in the Occupational Safety and Health Administration (OSHA) and the U.S. Environmental Protection Agency (EPA). Induction heating also can benefit from its close association with processes that have enjoyed good growth curves such as forging and other parts of the heat treating industry. For example, the forging industry, a vital factor in many industrial and commercial products, seems to be recession proof in the vision of many heating equipment manufacturers.

However, a strong growth trend for induction heating has not yet been demonstrated clearly. Preliminary statistics of the U.S. Department of Commerce for the 1977 Census of Manufacturers [42] actually show a decline in market share for some types of induction heating furnaces between 1972 and 1977. Out of all industrial furnaces and ovens there was a small advance in the market share of electrical furnaces, but the induction furnace market share fell from 16.4% to 13.5%. The prospect of large gain for a presumably better technology probably will not be realized until the anticipated natural gas price rise provides a parallel strong basic economic gain. Without gas price increases, the yield does not yet compare favorably enough with investment.

CHAPTER 4

LASERS AND ELECTRON BEAMS

During the past five years the use of laser and electron beams for metal heat treatment has advanced beyond the stage of a promising technical curiosity to a substantially well-accepted practice. The capital cost of laser or electron beam equipment generally runs over five times the cost for comparable induction heating equipment. This would be an insurmountable barrier if the choice rested solely on a cost comparison for comparable technical processes. However, the localization and control precision of laser and electron beams is so much higher that they provide qualitative advantages for the product that cannot be obtained by induction heating or any other means.

Essentially, lasers and electron beams introduce a thermal pulse into the work material with the same microscopic effect. Over a small region, the surface receives a high flux of heat energy through either the absorption of laser light by a surface coating or conversion of the kinetic energy of a stream of electrons bombarding the surface. For example, Cline and Anthony [43] analyzed the time-dependence of distribution of electrical temperature induced by a scanning laser or electron beam. Figure 4-1 illustrates the case of a laser beam melting a channel in the assumed x-direction. An electron beam of equivalent kinetic energy and scanning velocity would result in precisely the same heating effect. Results of the calculation in terms of instantaneous cooling rate and temperature distribution along x-axis [direction of scan for 304 stainless (thermal conductivity = $C_p D$ = 1/4 W/cm°C·sec)] are shown in Figure 4-2. The point source heat influx of 100 W and slow movement of 5 mm/sec are appropriate for melting a path of several millimeters on the surface, as would be desired in a welding application.

It is important to note how very close to the surface, less than 1 mm, the very high temperatures are confined. For the realistic case of a beam with a small spread, moving more rapidly to produce a hardening effect,

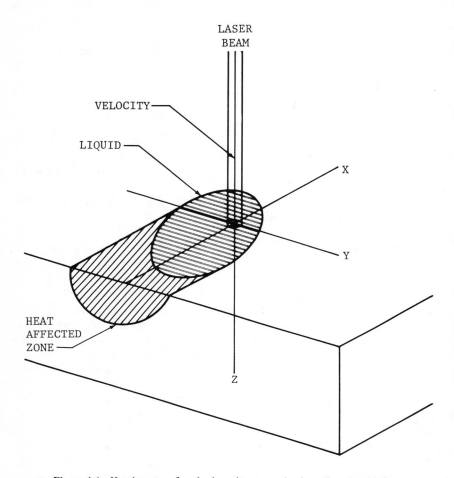

Figure 4-1. Heating at surface by laser beam moving in x direction [43].

for example, it is clear that both heating and quenching can be controlled together very precisely with no more elements to be handled other than the workpiece and the laser or electron beam. For exacting technical requirements, the localization and self-quenching attained with these processes are advantages, which repay the high capital cost of the equipment when the qualitative improvement of the product or service is marketable.

For almost all forms of metal-working, lasers and electron beams are direct competitors [44]. Electron beam (EB) systems are usually faster and more accurate because EB applies higher power levels and can be focused and guided using only electromagnetic fields, whereas lasers require

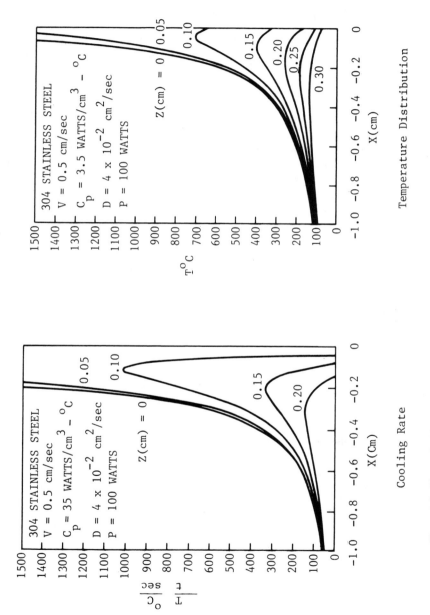

Figure 4-2. Theoretical heat transfer results for 100-watt point source scanning at 0.5 cm/sec [43].

mechanical adjustments in the optical path. However, electron paths can only be well-controlled in evacuated chambers and cannot be deflected as sharply or focused into as confined an area as lasers. There is a minor EB disadvantage in the elaborate arrangements that are required if a quenching medium must be used, whereas lasers can be used in that case without difficulty.

LASERS

The laser transforms light composed of a mixture of frequencies into coherent monochromatic radiation as the emission from stimulated transitions between molecular energy levels. In the example shown in Figure 4-3, input light acts to pump the population of molecular energy levels to produce a nonequilibrium excess at higher levels. If the electronic state lifetimes at these levels are relatively long, then a net excess of downward transitions is induced to the lower levels with emission of light corresponding to the transition energy difference. A coherent train of plane waves is developed by repeated reflection between precisely parallel mirrors at the ends of the chamber containing the molecules. One of the mirrors is partially

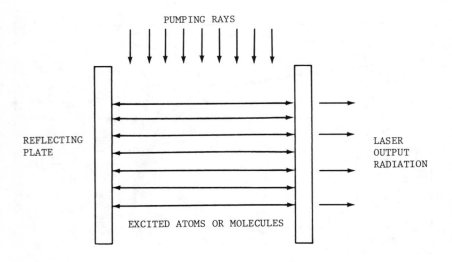

Figure 4-3. Parallel-plate laser.

transmitting so that the coherent monochromatic beam of plane waves emerges from it. This beam, because of the coherence, can be very narrow with minimal divergence so that the energy of the light is concentrated along a very confined channel. Recently, continuous lasers, based on the transitions among CO_2 molecular levels, have become available commercially with power output in a range of tens of kilowatts. These have become extremely important in metal heat treating applications.

Current types of lasers with their applications have recently been summarized by Spalding [45], as shown in Table 4-1. Costs for continuous CO_2 lasers over a representative group of manufacturers and power levels in 1976 are shown in Figure 4-4. From Table 4-2 it is apparent that the CO_2 laser, on a cost-per-watt basis, is the least expensive of the different types at $30–90 per watt versus $1000– 3000 per watt for argon-ion lasers. The figure shows the cost per watt for CO_2 lasers declining from $100 at the lowest power levels to less than $50 at approximately 20 kW. Some exemplary quotations from Laser Applications, Inc. [46] in Table 4-2 show that these costs have not changed substantially since then.

One way to relieve the high capital cost requirement is to time-share laser systems to work on several production lines [44]. The beam can be shifted so as to allow synchronized operation in sequence at separate locations, or it can be split into several beams so as to perform its work on several workpieces simultaneously. If several different processes are required on each piece, such as welding and heat treating, two stations—each devoted to one of the processes—can be powered by a single laser. Unique to laser beams, among electric metal treatment methods, the distances between separate output points may be as great as several hundred meters. Beams can be routed in simple sheet metal enclosures with little attenuation.

In addition to metal-working applications, lasers are used to drill and cut other materials such as plastics, rubber, cardboard, paper, cloth, ceramics and glass. They are also used as a diagnostic tool to measure, with very high precision, important physical phenomena taking place inside engines. For example, lasers are used to measure fluid flow and fuel droplet distribution [47], according to Doppler frequency shifts and light scattering effects. They are also used for very rapid noncontracting measurements of the pressures and stresses undergone by new alloy parts for advanced nuclear reactor systems [48]. Tube specimens made of six different experimental alloys and the Westinghouse Fast Flux Test Facility (FFTF) reference alloy, cold-worked 316 stainless steel, are inserted into capsules for testing in the FFTF. To measure the pressurized samples promptly after irradiation, enabling reinsertion in the next reactor operating cycle, a laser measurement was developed by E. R. Gilbert of Westinghouse. A laser beam illuminates the tube specimen and measurements are taken of the shadow to an accuracy of 0.5 micron.

Table 4-1. Some Typical Lasers and Illustrative Applications in Industry [45]

Laser	Wavelength (nm)	Mean Power (W)	Pulsed (J) and Repetition Rate (Hz)	Electrical Efficiency η (%)	Approximate Capital Cost 1976 ($1000s)	Typical Application
Argon	351-529	2-20	cw	<0.1	6-24	Thick-film microcircuit scribing
Nd-YAG	1,060	20-1,000 20-1,000	cw 0.5 J at 50 Hz	≤1 ≤1	12-80 ≥60	Cutting reactor wrappers Diamond drilling
CO	4,900-6,200	>100	cw > 1 kJ	40-60	–	Not used industrially
CO_2	10,600 10,600 10,600	~500 $(2-15) \times 10^3$ 400	cw cw 0.4 J at 1 kHz	~4 5-10 ≤10	44 ~400 23	Nonmetal cutting Welding metals Scribing ceramics

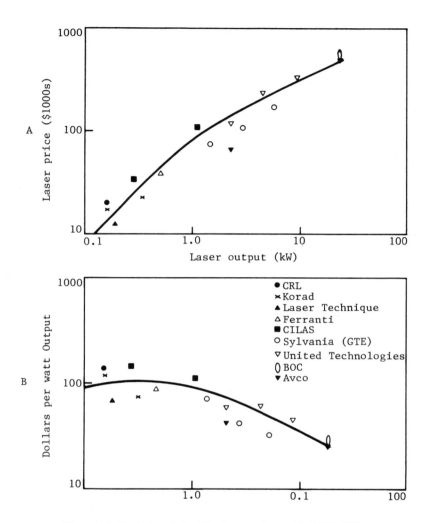

Figure 4-4. Capital costs for CO_2 laser equipment in 1976 [45].

Table 4-2. Recent Prices for Laser Systems [46]

Price	Installed Power	Production
$100,000	40 W	Small electric components
$250,000	2-3 kW	Industrial applications, hardening gears, etc.
$500,000	15 kW	Mass heat treating-camshafts, cylinders

In most aspects, lasers present no environmental problem or hazard to human health. However, there is potential damage to eyesight for operators of laser equipment. Various protective measures such as viewing windows, plate-glass filters, goggles and eyeglasses can attenuate either direct or scatter light. They can provide safety, in exposures of several thousand J/cm^2, when selected according to the wavelength of emitted laser light.

LASER MACHINING

In metal machining, precision optics and electronic circuits control closely the beam position and intensity to drill tiny holes, cut intricate parts, or mold delicate components. Since there is no physical contact with the workpiece, lasers cut virtually distortion-free parts. Mirrors and lenses can direct the beam around obstructions or into close spaces that are otherwise inaccessible. Since cutting is rapid enough there is no thermal distortion of the part or damage to adjacent heat-sensitive materials. The principal advantages of lasers lie in operations on very small and intricate parts. They can drill small holes in thin sheet metal for fuel nozzles and similar equipment where hole diameter and position are critical. They can make clean, neat, precision welds in delicate components such as thermocouples, strain gauges and watch parts.

Typical drilling and welding situations are illustrated in Figure 4-5 [44]. Laser-drilled holes usually have a 10% variation in diameter due to an entrance taper at the entry point of the beam and a thin layer of recast metal on the sidewalls. The beam is usually focused to maximum intensity on a region smaller than 2 mm diameter down to several microns. Incident power density is high enough to vaporize succeeding layers of metal forming a hole. For most metals the range of required power densities is 10^5–10^8 W/cm². These exceed the average power rating of most lasers so that it must be pulsed to obtain peak bursts of energy. The pulses are usually several microseconds long and repeated a few hundred times per second. For example, J. M. Langevin [49] of Gen-Tec, Inc., Quebec, has described a new carbon dioxide laser, T-CO2, which can drill holes in titanium and similar highly refractive metals without leaving a taper or appreciably disturbing the metallurgical characteristics of the adjacent material. The unit is made in sizes from 1–300 W and works with 1-μsec pulses repeated 300 times per second.

Another advantage of laser metal-working tools is that their structural members, manipulating a light beam, do not require the large mass and rigidity of conventional tools. This can be extremely valuable in new high-speed production technology, which includes, for example, cutting spindles rotating at speeds in the range 20,000 to 100,000 rpm [50]. Allen D. Gunderson, of Kearney & Trecker, pointed out that the main problem in developing

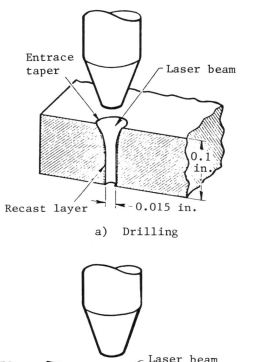

a) Drilling

b) Welding

Figure 4-5. Laser machining [44].

such equipment is that the steel structural members of the conventional tools do not have the strength and rigidity to withstand sharp changes in acceleration and deceleration involved in the high feed rates. Laser technology clearly can relieve this difficulty. However, to do it economically— for the raison d'être of high-speed technology is only cost savings—will require lower-cost energy supplies and more powerful lasers.

HARDENING AND OTHER HEAT TREATING

Lasers are used to harden selected areas on parts such as cams and valve seats where only small surface areas require treating. The powerful laser beam must scan the surface rapidly to avoid melting and spread the treatment path as shown, for example, in Figure 4-6. By dispersing the optical path, the intensity of the direct laser beam is reduced to about 100 kW/cm². In heat treating an outer bearing race, a 15-kW laser beam is dispersed and redirected to the surface by a toric focusing mirror with a workpiece feed rate of 4 mm/sec. The shaft is hardened in a similar way. Case depths are

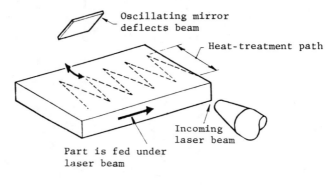

a)   General Principles

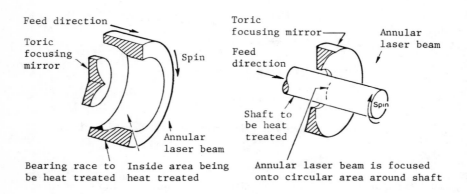

b)   Examples

Figure 4-6. Laser heat treating [44].

approximately 25 mm, with a Rockwell hardness index of 55. Surface temperature and depth of heat treating are controlled by exposure time and beam power. The process is self-quenching because such a small volume is heated that the surrounding cool material quickly conducts away the residual heat after the laser beam passes. The surface being treated is usually coated with colloidal graphite or similar light-absorbent medium. As the beam passes, the surface reaches the critical temperature for hardening and is then rapidly cooled by conduction to the cool interior. Frequently, a thin layer of alloying material is added to the base metal for special surface characteristics. Carbon and chromium added to low-alloy steel can extend high-temperature serviceability.

Wear resistance of interior surfaces of diesel cylinders has been increased by such laser treatment. When microscopic irregularities of metal surfaces touch due to the loss of oil film between them, as between the cylinder lining and piston ring, they form small "microwelds" that fracture under forced relative motion. This further roughens the sliding metal surfaces and the effect becomes cumulative. To provide a greatly hardened, more wear-resistant cylinder liner, they can be irradiated with high-power laser light for a self-quenched hard surface layer. Four 5-kW CO_2 lasers are now being used by the Electromotive Division of General Motors on liners for their 645 Series Diesel engines. The laser beam is directed down to the liner being treated from a second floor balcony. The work fixture rotates and elevates the liner following a prescribed program to ensure uniform heat treatment and to coordinate with related production processes. In Figure 4-7a the laser chamber itself is illustrated. The 5-kW lasers have been built to General Motors' specifications by GTE Sylvania's Western Division, Laser Products Group. A controlled mixture of helium, nitrogen and carbon dioxide gases is circulated through a glow discharge region maintained by a linear cathode and a set of anode segments. Mirrors at the ends of this region, one fully, the other partially reflective, cause the coherent laser light, a beam about 1.9 cm diameter, to emerge from the partially reflective mirror coated with a zinc selenide.

The liners are thereby case hardened to a depth of 0.25–0.64 mm in a process that is integrated into a high-volume manufacturing system and operated in the normal industrial environment. The laser heats only one small area at a time to a precisely controlled temperature for a precisely controlled time. The surface of the liner is painted black. Temperature of the overall metal piece remains low and no heat distortion of the piece occurs. The area under treatment cools rapidly and the proper microstructure is achieved for adequate hardness. A liner is treated in about 15 minutes but there are plans to reduce this to 8 or 9 minutes. More than 40 engines with laser-burned liners have undergone field tests for periods up to 36 months by railroad and marine customers.

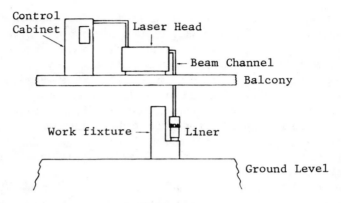

a) Production line arrangement

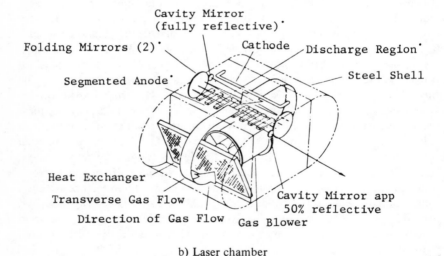

b) Laser chamber

Figure 4-7. Laser heat treating of diesel engine cylinder liners [51].

Lasers have been used to harden power steering gear housings at the Saginaw Steering Gear Division of General Motors since September 1974 [52]. About 15 CO_2 lasers (three at 0.5 kW and 12 at 1 kW) are used at three plant locations (two in Saginaw, Michigan and one in Athens, Alabama). Wear resistance of selectivity-hardened ferritic malleable castings is roughly ten times greater than those not treated. Distortion is essentially nonexistent because less than 28 g of the 6.3-kg housing is affected by the process of heating and self-quenching. Because of its structure, ferritic malleable iron is not easily hardened. The transformation hardening process by lasers had to be developed from scratch, since there was little prior experience with it in 1974.

The steering gear housings have been made with ferritic malleable iron (ASTM-A47-32510) since 1956. It is a good material for impact resistance, tensile strength, fatigue properties, noise-damping and machinability. However, in 1970, GM's Product Engineering Department foresaw potential increased wear in gear housings due to higher front-end loads by the extra equipment desired by customers and required by federal regulations. Extra hardness, such as that provided by a change of material, would reduce machinability. Machining is critical because a long bore had to be cut to very close tolerance in diameter and concentricity. Localized hardening, after machining without subsequent distortion, would solve the problem. Induction heating was tried, but it introduced distortion, which could only be relieved by extra heat treating. Laser heat treating proved to be the only economical one-step process.

The microstructure of ferritic malleable iron consists of temper carbon molecules in a ferrite matrix with less than 20% of pearlite. Since the available carbon is limited, hardening is difficult without redissolving the carbon modules and providing time for carbon to diffuse into the austenite matrix. In the laser hardening process, the surface temperature is controlled to $1038-1153°C$, and at a depth of 0.25–0.30 mm the temperature has fallen to around 810°C. Rapid dissolution and diffusion of the carbon modules takes place only very close to the surface. To absorb the $10.6-\mu$ laser energy the bores have been honed and coated with manganese phosphate, which enhances absorption to more than 90%. Laser beam power density is held to 38.5 kW/cm^2 within an incident circle on the surface of 1.8–2.0 mm diameter. The scan rate used is 2.5 cm/sec for 0.5-kW lasers and double that for 1-kW lasers. Depth of hardening is held to 0.03 mm in a thermal cycle of 30 seconds for a 0.5-kW beam and 18 seconds for a 1-kW beam.

In one of the most recent discussions, F. D. Siemens [53] of the IIT Research Institute in Chicago, has reviewed the merits of laser techniques to heat treat gears. He finds that production rate advantages tend to compensate for the extra energy costs of the laser systems. About 4 joules in the laser beam usually are required to get 1 joule into the workpiece to accomplish the heat treatment. The laser treatment seems to avoid requirement of later machining processes, such as grinding to restore surface quality. Such factors greatly influence process costs. However, there is evidence of an important quality advantage. The slip and wear character-istics of laser-hardened surfaces seem to be superior to those of induction-hardened surfaces when all parameters are otherwise the same.

ANNEALING SILICON ALLOYS

One of the most startling of recent laser heat treatment applications has been in the fabrication of silicon-based circuits, both for microprocessors and photovoltaic devices [54].

Laser processing for silicon had been explored in the U.S. very early [55], but it was later work in the U.S.S.R. around 1976 that sparked the recent surge in U.S. work. The drive in microprocessor development has been toward more compact arrangement of circuit elements on a single silicon chip. This has required more finely delineated regions of alternating conductivity and shorter paths for charge conveyance. Direct implantation of impurity elements by high-energy beams has become the best way to achieve well-controlled concentrations and spatial configurations. The ions need not penetrate far into the silicon, since all the important charge transfer of the circuit occurs within the first micron surface layer. However, the high-impact energy of the ions tends to disorder the parent silicon crystal lattice. Annealing the disorder by heating the entire device allows new contaminants to enter the lattice, some diffusion of the implanted ions, and some dimen-sional changes. All of these effects degrade the circuit performance. In laser annealing, the required heating remains localized with a temperature high enough to permit both impurity and host atoms to return to normal crystal lattice positions. Improvement of laser annealing on depth localization of impurity atoms is illustrated in Figure 4-8 [56].

Generally, the lasers used have been of lower power, such as the ruby laser used by a group at Oak Ridge National Laboratory, which impacts pulses of 1 J/cm^2 in 50 nsec. This is enough to reduce the size of defect region, due to ion-implanted boron, phosphorus, arsenic and antimony, to less than 10 Å. However, there is some diffusion of the implanted impurity atom from these original intended regions. By raising laser beam power and shortening the pulse time to 15 nsec, the Oak Ridge group has been able to raise impurity concentrations and reduce diffusion from the implantation site [57]. According to C. C. White of the group, an impurity concentration far in excess of equilibrium solubility limits can be achieved. The lasers essen-tially melt the surface layer and it recrystallizes in approximately 100 nsec, freezing the impurity atoms in their proper lattice sites and eliminating surface defects.

A CW laser has been used by groups at Bell Telephone Laboratories, Stanford University, University of Illinois and at the Advanced Research

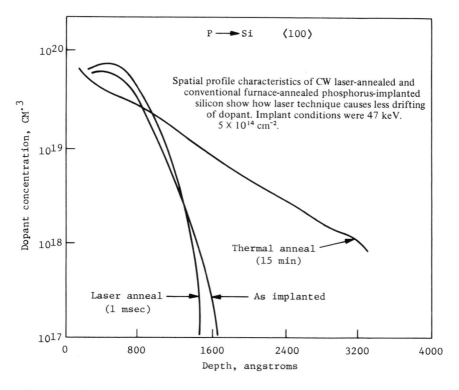

Figure 4-8. Comparison of laser and thermal anneal on depth diffusion of implanted impurities [56].

and Applications Corporation in Sunnyvale, California. Using a scanning procedure, they have been able to avoid surface melting and thereby prevent any impurity diffusion at all. This has distinct advantages for implantation of very closely spread circuits on a single silicon chip. To date, current industrial practice has not otherwise achieved the required precision control.

The utility of lasers in annealing photovoltaic devices is owed mostly to the speed of the process compared to the slower thermal annealing. This is because of the high premium placed on reducing the production cost of photovoltaic cells, which can be made in very large amounts to provide significant energy sources. As with microprocessor circuits, the annealing requirements originate in the preference for ion implantation of impurities needed to establish the photovoltaic junction properties. Silicon wafers 7–10 cm diameter have received impurities on their entire surface by ion implantation to establish a sharp vertical fall-off in distribution. Furnace annealing of the crystal structure in a batch process would reduce the vertical impurity gradient and take up to two hours. Laser annealing can be incorporated in an automated sequence with ion implantation for a much faster production rate. The Spire Corporation of Bedford, Massachusetts has been developing such a facility to manufacture one acre of cells per day. However, it has substituted a pulsed electron beam for the CW laser in the annealing stage.

ELECTRON BEAM HEAT TREATMENT

Electron beam systems for heat treatment have been characterized by Dreger [58] as faster than laser systems, more accurate than induction heaters, more selective than flame hardening, and far more energy efficient than all three. They are, in fact, most similar to the laser system as indicated above, sharing an almost identical set of industrial applications. The driving advantages over lasers are that electron beam systems are available with much higher power outputs, at lower capital costs, and at a much higher utilization of power supply energy.

The entire electron beam system, complete with vacuum pumping system and work chamber, either a beam programmer or a minicomputer control for precision beam placement, and interface hardware and software, typically ranges in cost from $200,000 to $500,000. Auxiliary costs equal to these for accessories to provide multiaxis workpiece manipulation and automatic load-unload capability would be the same as for lasers. The EB system up to the beam output may be only one-third to one-fourth the cost of a comparable laser system up to the laser beam output. Moreover, while present commercial lasers are limited to 10-kW beams at best, there are EB systems that can put up to 500 kW into the work. In converting input power to light, lasers realize only 7–10% in the beam. For EB systems, about 75% of input power is placed in the electron beam. This is better than high-frequency induction heaters at 50%.

A typical EB system using a partially evacuated work chamber is illustrated in Figure 4-9 [59]. Electrons are emitted from a heated cathode (located inside the part labeled electron gun), which is held at a negative potential. The cathode is shaped so as to focus most of the electrons into a beam. The electrons are drawn toward the grounded anode, which contains an aperture through which most of them pass toward the magnetic lens. This lens, in turn, reshapes the beam to refine it as required for operation in the partially evacuated work chamber.

Since enough of the electrons in the beam are deflected by air molecules to substantially reduce beam intensity, there is a basic economy in the use

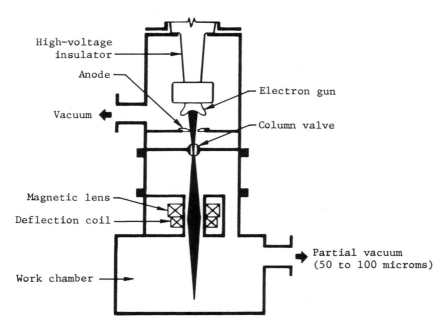

Figure 4-9. Electron beam system operating in a partial vacuum [59].

of evacuated work chambers. However, they are not essential if the work can be placed very close to a beam exit nozzle, which must also function as a partial air lock to protect the vacuum needed by the operation of the gun itself. Although EB guns operate at the low pressure of 0.1 μ, the work chamber in vacuum EB systems may be maintained above 50 μ. Such a working level vacuum can be reached quickly by mechanical pumps (5 seconds for a 28-liter chamber). The beam of electrons may be controlled by both the magnetic lens and the electric potential applied between cathode and anode.

Power loss due to working in air can be held to an acceptable level by close placement of the work. Capability to sweep the beam is reduced so that the work must be waved or rotated instead. To raise the intensity of the beam, the applied voltage may be increased, but this can produce hard X-rays from the gun's interior so the gun must then be shielded.

There are two companies in the U.S. that build EB heat treating systems. Those of Skiaky Bros., Inc., in Chicago include a vacuum workchamber and operate with electron accelerating voltages between 50 and 60 kV. The beam current is approximately 750 mA, variable according to requirements, to put up to 45 kW into the beam incident on the workpiece. A steel or stainless steel enclosure is used for shielding. Leybold-Heraeus Vacuum Systems, Inc., of Enfield, Connecticut builds nonvacuum and partial-vacuum systems. The nonvacuum system works at a 30-kW beam power level and requires lead shielding. The partial-vacuum system uses only 25-kV accelerating voltage to put 60 kW of beam power into the workpiece.

With the beam sharply focused for maximum power density, it typically induces melting entirely through the workpiece to operate as a cutting or machining tool. At lower beam densities, the beam will melt the surface of the workpiece to be used for welding. Its early successful use for welding provided the base from which EB technology developed into other applications. With the beam defocused further, it can be used to heat the surface without melting to various temperatures useful in hardening or annealing according to the dwell time of the beam. Typical heat treatment beam distributions and beam scanning patterns to cover extended surface areas are shown in Figure 4-10.

Since irregularities in the electron emission source will be reflected as irregularities in the intensity distribution of a defocused beam, the beam must be swept over a surface being treated to heat it uniformly. There are some EB systems designed specifically for heat treating that can produce a large uniform spot. Usually, though, an EB system is generalized, and to include the welding capacity will use a high-intensity electron source that requires beam scanning for uniformity. The Leybold-Heraeus EB system moves the beam continuously over raster patterns such as those shown in Figure 4-10b. The EB systems of Skiaky Bros. move the beam in discrete steps over a pattern of dots arranged like the raster patterns. The time for a step movement is about 10 μsec. In normal operation, a dwell time of 20 μsec is used at each dot. The dots are spaced at 0.125–0.500 mm, well under the typical beam distribution width of 3 mm shown in Figure 4-12a. Beam travel over a workpiece surface for both Leybold-Heraeus and Skiaky systems is usually done at 10 meter/sec so that a specific area could be exposed repeatedly about 150–200 times/sec. To assure self-quenching, the thickness of carbon steel and low alloy materials should be at least four times the depth of the hardened core desired (usually somewhat over 0.1 mm). Thinner parts may be placed on auxiliary-cooled base plates to assure their self-quenching. Some comments on recent experience with specific EB applications are discussed in the following subsections.

HARDENING

Recently, Terry R. Gonser [12], of GM's Detroit Diesel Allison Division, compared the Skiaky and Leybold-Heraeus systems for GM's hardening applications on a production line. In this case, the choice was Skiaky because

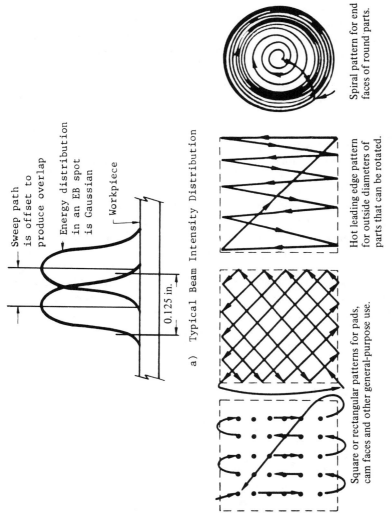

Sweep path is offset to produce overlap

Energy distribution in an EB spot is Gaussian

Workpiece

0.125 in.

a) Typical Beam Intensity Distribution

Square or rectangular patterns for pads, cam faces and other general-purpose use.

Hot leading edge pattern for outside diameters of parts that can be rotated.

Spiral pattern for end faces of round parts.

b) Typical EB Raster Patterns

Figure 4-10. Electron beam energy input [58].

the computerized control allowed the highest production efficiency. Leybold-Heraeus intend to, but have not yet, developed the same automated control. Capital investment was $460,000 and it was 8 months from order placement to delivery and setup. The system was bought to harden the ball seat surface of a valve cam lifter. This area is 3.8 cm down the interior of a cylinder. Induction heating coils were difficult to design for it. Both lasers and electron beams were appropriate. However, the laser needed a black surface coating, which proved expensive, its energy coupling to the work was low enough that energy costs were significant, and quality control measures seemed costly.

The EB system provided better control with lower system cost. Use of the vacuum chamber was no detriment. Parts are loaded about 4 at a time in the 28-liter work chamber, which is pumped down to 40 μ (10-5 atm) in 3–5 seconds. The gun chamber vacuum of 0.1 μ, requiring a diffusion pump, is maintained separately. Energy coupling efficiency of beam into work is 97%. Gross production rate is 626/hr. Computer control of the beam, extremely important in the production process, is faster and more precise than in GM's laser installation, in which the beam must be shifted by mechanical mirror rotation. The Skiaky system was first described by C. L. Gilbert [60], chief application engineer at Skiaky. He pointed out that in addition to control of beam illumination and average intensity, with time the energy profile of the beam pulse also can be set. In the typical case, more power is required when the part is cold. The beam current, for example, can start up 260 mA and be set to decay exponentially to 100 mA. Within the Skiaky electron gun there is a control electrode, just outside the filament cathode, which can be set to adjust the beam voltage-current characteristic with a small bias voltage. Typical operating parameters for the Skiaky EB system are shown in Table 4-3.

CORROSION-RESISTANT COATINGS

The electron beam concept has been used to apply physical vapor-deposited (PVD) coatings to protect gas turbine parts from high-temperature corrosion. The coatings are usually based on a combination of Cr, Al, Y with Co, Ni or Fe [61]. The metallic coating ingredients are bombarded by one or more electron beams to be vaporized [62]. This vapor then condenses on the turbine blades, which are also in the vacuum chamber, and the coating thickness can be controlled precisely by the time of exposure and the speed and angle at which the parts are rotated in the vapor. One example is the latest development in the application of PVD coatings to aircraft gas turbines, subject to operating temperatures of 980–1150°C, an evaluation of Co, Cr, Al, Y and Ni, Cr, Al, Y coatings for the Pratt and Whitney JT8D engine, which is now the most widely used in the world. In Table 4-4, a preliminary assessment of various coatings shows that a PVD Co Cr Al Y coating will increase the blade life by a factor of five over presently used aluminide-coated blades.

Marine gas turbines operate at much lower temperatures, 700–870°C, but in a high salt spray atmosphere. The early General Electric LM 2500 engines, used in the U.S. Navy's DD-963 destroyer, had an aluminide coating on the first-stage blades providing engine lives of only 2000 hours. The blades were being corroded by a sulfidation/oxidation combination caused by the presence of sodium sulfate. Changing the blade superalloy and using the

Table 4-3. Operating Parameters and Normal Tolerance
for Electron Beam Treatment [60]

Parameter	Operating Range	Tolerance
Applied Potential	6–60 kV	1% or 1 kV
Current	0–700 mA	1% or 1 kV
Energy Spot Location	Circle of 0.95 cm diameter	0.075 mm
Spot On Time	99 sec	0.1 sec
Focus Current Density	0–63 amps/cm^2	0.1%
Vacuum	0.1–100 μ	10%
Part Position	Variable	± 25 μ

Table 4-4. Preliminary Assessment of Protection of JT8D Blades from Sulfidation
Corrosion (based on laboratory and engine evaluations) [61]

Type of Coating	Blade Material	Coating Life Improvement Factor	Base Metal Life Improvement Factor
PWA 68 CoCrAlY+	IN 792	5+	4.5
PWA 268 CoCrAlY− PWA 268 over	IN 792	4+	4.5
PWA 70 COCrAlY−over Or Diffusion PWA 270 over	PWA 1455	4+	1.0
PWA 273 NiCoCrAlY over Aluminide	PWA 1455	4	1.0
BB Rhodium Aluminide	PWA 1455	2+	1.0
PWA 73H Pack Aluminide	IN 792	1.5	4.5
PWA 73L Pack Aluminide	PWA 1455	1.0	1.0

electron beam PVD Co Cr Al Y increased the life of the first-stage blade to
4000 hours. More recent coatings of higher Cr and lower Y promise to extend
the life to 10,000 hours.

For industrial gas turbines, which operate in the 760°C to 930°C range,
a very large 200-kW electron beam coater has been developed for Saudi
Arabia. This coater handles parts weighing 41 kg, with dimensions of half
a meter on each side. In addition, the components are multivaned. Saudi
Arabia uses high sulfur oil-burning for the turbines so they would be subject
to a similar sulfidation/oxidation. The standard Co Cr Al Y PVD coating
on blades already run for 13,000–15,000 hours promises a protection life
of 200,000 hours.

To indicate the range of possibilities in further coating developments,
Figure 4-11 shows the temperature difference between melting and boiling

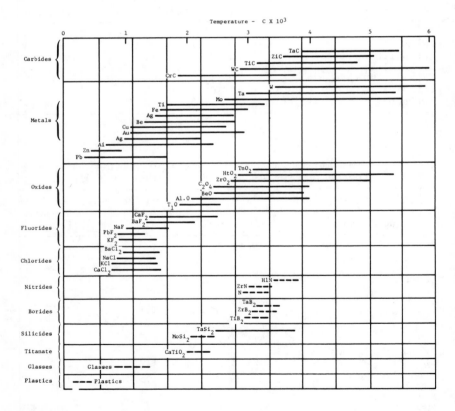

Figure 4-11. Approximate melting point and boiling point range of physical vapor-
deposited (PVD) materials applicable for turbine coatings.

point of the constituent materials. These materials have all been physically vapor-deposited by the electron beam process.

WELDING

Electron beam welding applications have been reviewed by Marcelin [63]. In general, these have continued along the lines established at the start of EB metal work. However, a particularly striking development has been described by Kelly [64] as an integral operation of the Wallace Barnes Steel computerized rolling mill in Bristol, Connecticut. Two electron beam units are used to produce continuous bimetallic strips. They each have beams of 1.5 kW power, 0.1 mm diameter, projected into a hard vacuum chamber (less than 0.1 μ). In operation, two coils of strips feed continuously into the vacuum chamber at speeds of 13–20 cm/sec. An adjustable guidance system mates the strips and horizontally aligns them at the point of beam incidence. The single-welded bimetallic strip then passes out of the chamber and is wound onto a takeup reel. Typically, a thin strip of hard, abrasion-resistant tool steel is joined to a tough, less expensive backing steel with good fatigue properties. Such a bimetallic strip is used for hand saw blades, power hacksaws and other tools requiring a cutting edge that can flex continuously over pulleys at high speeds. Such blades also better withstand the thermal shocks and impacts of metal cutting operations.

CHAPTER 5

ELECTRIC ARC AND PLASMA PROCESSES

Since 1976, electric energy use to carry out the prodigious heating required to produce primary metals has become economic by a combination of fossil fuel price increases, the environmental and production control problems posed by fossil fuel burning, and the unmatched control with which electric supply methods can be managed. On this issue, the use of arc and plasma furnaces is especially pertinent since at recent conferences the most exciting technical issues surrounding these concerned automation and control procedures they made possible. A combination of productivity gain and worker safety was clear and proven justification for the substantial costs and development efforts incurred in automation programs.

While technically similar, electric arcs are distinguished from plasma in that their use has begun a clear program toward dominance in the economically massive steel-making industry. It has been pointed out that on a worldwide basis the portion of steel-making contributed by electric furnaces has risen from 13% in 1970 to 19% in 1980. Therefore, most of this section emphasizes fairly detailed discussions of the most recent electric arc furnace developments presented at U.S. conferences.

According to estimates within the industry, the total U.S. production by electric steel-making furnaces (mostly arc and submerged arc) was to surpass 30 million metric tons in 1979. The annual production rate has been growing at 10% each year recently, while all U.S. steel production is virtually static at around 110 million metric tons. A major reason is the contribution electric arc furnaces make toward productivity. The following subsections outline briefly the remarks by a number of industrial leaders to illustrate their general philosophy on the issues currently being raised.

JONES AND LAUGHLIN ELECTRIC STEEL-MAKING

To provide a specific example of how to take maximum advantage of electric furnace capability, Thomas C. Graham [65], President of Jones & Laughlin, has described productivity improvement steps taken at their electric furnace shops, which are among the largest in the U.S. Citing the current capital shortage in U.S. industry, he said that after the investment for equipment has been made the company must focus strongly on the personnel operating and managing it. A climate to favor development of innovative ideas can be created by an organizational structure that includes elements such as a strong program in personnel development, incentives for individual initiative, and channels of communication for further exchange of ideas between operational and management personnel. This is particularly important when the goal is productivity improvement without new capital expenditures.

As an example, Graham cited the new J. and L. electric furnace shop, which was a major capital investment from which they desperately needed to exact a maximum return. Two large electric arc furnaces were built in the middle of an old open hearth shop in their Pittsburgh plant. Each of the furnaces draws 110 MW to melt 250 metric ton batches and to produce about 140,000 metric tons annually. They have water-cooled side panels and use electrodes 5.3 meters long and 0.71 meters in diameter. The furnace crucibles are supplied and tapped with 6.2-meter-diameter buckets operated by an automatic ladling system.

Personnel for the new furnaces were drawn exclusively from open-hearth crews. For well over a year before the furnace began operation the maintenance and operation crews were given special training. This included a five-day visit to a J. and L. electric shop already operating in Cleveland. A few months before operation began the crews practiced dry runs to familiarize themselves with the controls and the operating cycle. The faster cycle of the arc furnace requires much more extensive coordination. Since beginning operation, supervisors have held periodic meetings with furnace crews to discuss operations and possible procedural changes. Graham claims that, overall, the new furnaces, in addition to a drastic reduction of the environmental impact of the open-hearth furnace, have produced steel ingots at substantially lower costs.

A NATIONAL VIEW OF PRODUCTIVITY

Ira Greggerman [66], of the American Productivity Center in Houston, Texas, reviewed national characteristics of the U.S. decline in productivity to indicate the most likely productivity improvement measures he would recommend on a national basis.

Using data on the output per man-hour and U.S. trade balance in two areas, comparing the rising productivity in 1947–1967 with static productivity in 1967–1979, Greggerman drew the clear implication that a decline in productivity implies increased unit labor costs in a repetitive cycle:

Greggerman contends that a forced increase in productivity essentially can reverse the cycle. The major features of national economic life, which he feels are important in promoting activity through specific measures, are as follows:

- Government—Estimates by Chase Manhattan are that about $100 billion would be needed to bring U.S. manufacturing plants up to the standards of West Germany and Japan. U.S. plants are 18 years old on average, while those of West Germany and Japan are 10 years old.
- Capital—Because there are no tax incentives for investment, U.S. individual savings rates are less than 7%, compared to West Germany at 25% and Japan at 33%.
- Research and Development—Since 1967, measured in constant dollars, there has been no advance at all despite continuing rise in gross national product. This has resulted in a slowdown of new technology development.
- Energy Requirements in Manufacturing—Per unit of production the ratio of energy cost to labor cost has been rising back up to the levels that held before World War II.
- Human Resources—Greater stress urged in proper development of people. The physical assets make things possible but people make them happen.

STEEL INDUSTRY VIEW OF PRODUCTIVITY

Alvin L. Hillegas [67], vice president and general manager of the Eastern Division, U.S. Steel Corporation, reviewed the concerns of the steel industry. In the recent past he has been able to cite capital improvements to a 25-year-old rolling mill, which raised its annual productivity from 2 to 2.7 million metric tons. Hillegas stressed that there were also overall management issues pertinent to productivity improvement that should be aired.

Table 5-1. Crude Steel Production Compared to Population
(millions of metric tons and millions of people) [68]

	1967	1968	1969	1970	1971	1972	1973	1974	1975	1976	1976 Population
World	493.6	528.7	572.5	593.9	579.7	626.4	693.5	704.3	643.8	675.0	
West Germany	36.7	41.2	45.3	45.0	40.3	43.7	49.5	53.2	40.4	42.4	61.5
Japan	62.1	66.9	82.2	93.3	88.6	96.9	119.3	117.1	102.3	107.4	112.8
France	19.7	20.4	22.5	23.8	22.9	24.0	25.3	27.0	21.5	23.2	52.9
United Kingdom	24.3	26.3	26.8	28.3	24.1	25.3	26.6	22.3	20.1	22.3	55.9
U.S.S.R.	102.2	106.5	110.3	115.9	120.7	125.6	131.5	136.2	141.3	144.8	256.7
Italy	15.9	17.0	16.4	17.3	17.5	19.8	21.0	23.8	21.8	23.4	56.2
U.S.	115.4	119.3	128.1	119.3	109.3	120.9	136.8	132.2	105.8	116.1	215.1

For many years the steel industry has centered its attention on total production rather than production per man-hour. This attitude was acceptable so long as the U.S. steel industry was dominant in international trade, as it was in the early 1960s and before. However, by the mid 1970s annual production from both Japan and the U.S.S.R. had reached a par with the U.S. and the U.S. steel industry clearly would have to improve its labor productivity if it were to compete in the international market. As shown in Table 5-1, none of the major steel-producing countries, except the U.S.S.R., has a growing rate of steel production throughout the 1967–1976 era. By 1976, except for West Germany and Japan, all these nations had total steel production roughly in proportion to their population.

Hillegas emphasized that labor productivity would have to be the key to revive growth in the U.S. steel industry. Traditionally, capital investment would be the quickest and most efficient path to productivity increase. However, as a whole, the U.S. steel industry was woefully short of the capital required. For improvement in utilization of labor he recommended the following as major points:

1. More flexibility in labor and craft practice was needed with respect to functional divisions and workshop organization.
2. Better system of incentives was required to provide strong motivation.
3. More understanding by management of what is essential in the work force was needed. Hillegas decried some of the prejudices against "hippy-like" appearance, which he was not able to correlate with poor performance.
4. Industry-wide conferences should be held to provide forums in which a broad spectrum of opinions could be exchanged among labor and management representatives.

ECONOMIC ASPECTS OF BUILDING AND SUPPLYING POWER TO NEW ELECTRIC ARC FURNACES

William Lubbeck [69] of Union Carbide pointed out the difficult relation between utilities and steel-making companies installing new arc furnaces with their high power demands. There are many electric utility companies that lack the available short-circuit capacity or systemwide stability needed to deliver electric energy at the MVA flowrate the melt shop requires. Ideally, the available power (short-circuit capacity) should be 35 times the rated capacity of one furnace transformer. This is alleviated somewhat by the fact that in multifurnace shops not all furnaces are running at full power all of the time. When the short-circuit capability of the electric utility company falls short of the requirement of the arc furnace, the supply voltage under full load operation drops below the threshold of acceptability, the arc length changes and the melting efficiency of the furnace declines. This means a reduction in productivity for the melt shop and reduced dollar revenue for the utility.

The appropriate solution requires that a detailed power system analysis be undertaken by the steel company and the electric utility company so that the productive capability of the melt shop can be matched realistically to the capability of the utility company to furnish an adequate flow of power. There is a prevalent lack of appreciation of the four-way interdependent relationship between the arc furnace electrical load factor, energy cost, demand cost and productivity. For an arc furnace to operate most efficiently it must be run at full power and full capacity for as long a period as economically feasible during a heat. The time utilization factor, percentage power-on time, ideally should average 70% of the total heat time.

An effective tool for limiting arc melting power costs is well-designed power demand control. In a high-powered arc melt shop the power bill alone will be about 25–30% of the total cost of converting raw materials to molten metal. Demand time intervals are negotiated with the utility company and their length is extremely important. Since energy consumption is nearly constant and only increases in line with increased melting interruptions, it follows that the longer the demand time intervals and the fewer interruptions, the more efficient the production operation.

It would be a mistake to count only the time involved in the actual demand interruption. Actually, the losses in energy and, consequently, the increase in conversion cost are much greater due to the radiated losses of energy from the furnace during the delay period. Since the operator has a greater amount of energy at his disposal with a longer demand time interval, interruptions consequently are reduced. For the benefit of both the melt shop and the power company, a high production shop should not be exposed to excessive demand shutdowns or any other delays. Power and operating procedures vary in melt shops depending on the product made and type of equipment used. Every arc furnace has its own electrical and melting characteristic and seldom can be compared with another furnace. The efficient supply of power to the arc furnace melt shop requires a continuing dialog between the electric utility company and the melt shop.

In this connection, John A. Galbreath [70] of the Whiting Corporation discussed the special problems of small electric arc furnaces in what have become known as minimills. A minimill may be described as a relatively small producer of steel products, approximately 50,000–500,000 ton/yr, that can compete with major steel producers by minimizing the capital investment and manpower required to produce steel products, and by being capable of delivering, on short notice and at reduced freight costs, small quantities of these products within a relatively short distance from the plant. It is necessary that the mill locate in an area where raw materials, scrap, electricity, water, transportation and manpower are economically available.

The original minimills started to appear on the United States scene somewhere in the mid 1950s, using electric arc furnaces of approximately 5- to 10-ton capacity, and annual production rates in the range of 40,000 to 80,000 tons of finished product per year. They were used primarily for producing reinforcing bars for the highway and small construction industries. At that time the capital investment of such a minimill was less than $50 per annual net ton of capacity.

By the 1960s arc furnace sizes in minimills ranged from 50- to 100-ton capacity with annual output up to 150,000 tons. Around 1970 existing furnaces were retrofitted with much larger power supplies—typically doubling the rate of electric energy input. The 20-foot-diameter furnace of 100-ton capacity was now supplied through a 60-MVA transformer, compared to 30 MVA previously. Most recently, such a furnace started operation at Raritan River Steel in Perth Amboy, New Jersey, with a 93-MVA transformer. At peak input it operated at 105 MVA and supplied energy at the rate of 70–80 MW. Annual capacity is 500,000 ton/yr of finished product. This output level is only exceeded by the largest major steel company furnaces operating at several hundred MVA. The published figures on Raritan River Steel state total investment for the new mill at $94 million, or $188 per annual ton of capacity.

At the present time there are more than 60 minimills in the United States with annual nominal capacity of more than 15 million tons, and all present indications tend toward continuous growth of minimills with arc furnaces of larger size and greater production capability, as the prime melters. From its inception the minimill has used arc furnaces for producing its molten metal requirements. The arc furnace has the capacity of being able to melt economically the cheapest grades of scrap without costly preparations, and has the flexibility to be turned "off" or "on" on short notice, depending on demand for steel products.

A typical arc furnace has a power line in the 15-kV or 34.5-kV class, feeding the furnace transformer through a set of furnace disconnects and a furnace switch, usually a vacuum switch. The furnace transformer steps down the voltage to furnace voltage. Usual ranges are approximately 250 V phase-to-phase under no-load conditions for 8- to 10-ton furnaces, and up to 800 V and higher for 400-ton furnaces. Secondary currents range from 10,000 A on smaller furnaces up to 100,000 A and higher on the large furnaces. The furnace secondary conductor design, usually consisting of water-cooled copper flexible cables and buswork, is very critical since operating at these high currents can cause undesirable voltage drops and imbalance in the phase impedances, resulting in an imbalance of power to each arc.

A typical melt cycle for an arc furnace in a minimill can be broken down into three main parts: meltdown, superheat and refine, and power off for tapping, charging, etc. Meltdown takes place as the three electrodes bore holes down through the scrap, the holes being of sufficient size to ensure that a molten pool covers the bottom so that arcing on the bottom refractory does not occur. As arcing continues at maximum power input on the molten pool, the scrap slides into the pool and the charge literally melts from the bottom up. When sufficient melting has taken place to allow for a back-charge, a second charge is dropped into the furnace and the meltdown procedure starts again. As many as five or six backcharges have been used because of typical low-cost scrap density and the economic reality that the larger the tonnage poured for a given size furnace, the more economical it becomes.

After all the scrap has been melted down, it is necessary to reduce the level of power input to the furnace to avoid damage to the sidewall refractories. The furnace now is in what is called a flat-bath condition and the superheat and refine period starts. Generally, the level of power input during this period is approximately one-half to three-quarters of meltdown power. Samples are taken and the temperature of the molten metal is measured. When desired analysis and temperature are reached, the heat is ready to tap. Power is turned off and the furnace is tilted to pour the molten metal into a transfer ladle. The furnace is then checked for refractory erosion and, if no patching is required, is charged for the next heat cycle. Typical electric energy costs for an arc furnace used for producing molten steel in a minimill are approximately $15/ton poured. (This is based on an energy consumption of 500 kWh/ton and a cost of $0.03/kWh.) A typical new minimill with 500,000-ton annual capacity is then worth $7.5 million annual revenue to the supplying utility.

The utility company will need to be able to supply electric energy at high rates to properly feed the arc furnace in question. For most furnace installations the power supply should have a short-circuit capacity at least 50 times and, preferably, 100 times the nominal capacity of the furnace transformer. There is also the mill need to obtain the maximum power level to the arc furnace. This requires that all the power lines and stepdown equipment on the lines feeding the arc furnace have the minimum practical impedance. A final consideration relating to electric energy costs for arc furnaces is demand billing. The arc furnace is not a steady load on a system. Meltdown is at maximum power, refining is at approximately one-half maximum power and the power is off for charging, tapping and other nonprocessing operations. Hence, available demand never is utilized fully. Multiple furnace loads with some type of demand limit system can more fully utilize the available demand. However, due to the inherent characteristics of the arc furnace melt cycle, the furnace loading cannot always be increased or decreased to fully utilize the established power demand.

EXPERIMENTS IN ARC FURNACE OPERATION

Because of the markedly heterogeneous nature of the furnace charge and the high demand placed on the electric supply system, which results in unusually high interaction between load and generator, the detailed science of how arc furnaces operate is still being developed. Some of the most recent developmental experiments aimed at settling some of the most important issues are reviewed in the subsections below.

Excavation of a 75-MVA High-Carbon Ferromanganese Electric Smelting Furnace

A large cooperative project was organized in South Africa to gain some idea of the spatial distribution of melt constituents in an arc furnace by excavation of a furnace that has been allowed to cool. Results were reported by Nicholas Barcza of the National Institute for Metallurgy in Randburg, South Africa, Andrew Koursaris of the University of Witwatersrand in Milner Park, South Africa, William Gericke of Metalloys Limited in Meyerton, South Africa, and J. Bruce Lee of the University of Nevada in Reno, Nevada [71].

An opportunity was presented by a planned shutdown of a 75-MVA high-carbon ferromanganese electric smelting furnace in October, 1977 at a plant of Metalloys Limited in South Africa. A team of metallurgical engineers, research students and technicians from the South Africa National Institute of Metallurgy and the University of Witwatersrand sampled the furnace contents during its "dig out" in an archeological manner. Approximately 75 samples were taken from positions at the top, sides, interior and base of the furnace. Figure 5-1 shows the furnace with the sample locations coded.

There were a number of zones in the fused melt with distinctive characteristics that could be related to the metallurgical processes that had been taking place. First there was a zone close to the electrode. This was surrounded by an outlying annular zone stretching to the furnace wall and then, underlying the electrodes, a series of layers. The zones are outlined in Figure 5-2 as occurring in the section AB of Figure 5-1. Sintered material surrounds the electrode above the presumed main electric conduction paths through the furnace material constituents. Below the electrode the layers, in sequence, are coke-enriched, manganese oxide melt and ferromanganese alloy at the base.

Raw feedstock materials in the burden consisted of manganese ore, quartzite and a coke-coal mixture for carbonaceous reduction. Radioactive tracer studies with Fe-59 showed that the burden descends rapidly in the regions surrounding the electrode to appear loosely centered. In the outlying regions 2 and 3, the mix is partly reacted with carbonaceous enrichment toward the furnace walls. Beneath the electrodes, feed material components segregate as smelting takes place. The reducing coke forms

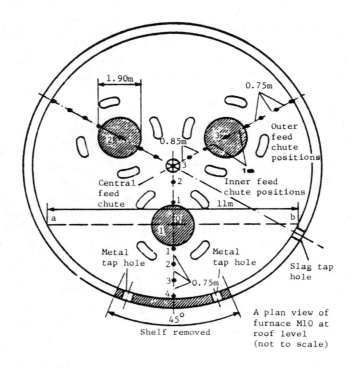

1.90m

0.75m

Outer feed chute positions

0.85m

Central feed chute

Inner feed chute positions

a

11m

b

Metal tap hole

Metal tap hole

Slag tap hole

0.75m

45°

Shelf removed

A plan view of furnace M10 at roof level (not to scale)

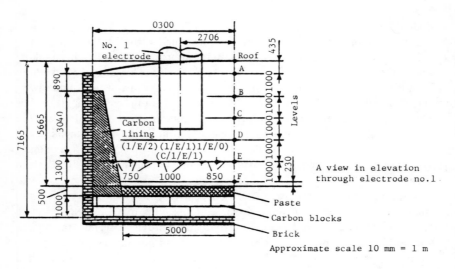

0300

2706

435

No. 1 electrode

Roof

A

890

B

C

5665

3040

Carbon lining

D

7165

(1/E/2)(1/E/1)1/E/0)
(C/1/E/1)

E

Levels

1300

750   1000   850

F

230

500

1000

5000

Paste

Carbon blocks

Brick

A view in elevation through electrode no.1

Approximate scale 10 mm = 1 m

Figure 5-1. Selected sampling locations for the arc furnace "dig out" [71].

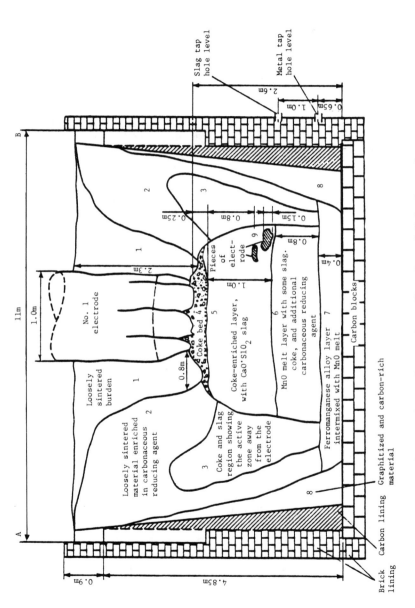

Figure 5-2. Locations and approximate dimensions of distinctive arc furnace zones determined by the "dig out" [71].

a bed just below the electrode in region 4, followed by a crumbly coke-enriched slag layer in region 5. The layer of MnO-melt that forms in the next region, 6, was predicted by some independent mineralogical studies. Grains of almost pure MnO were observed to form from the heating and partial reduction of the manganese ore. It has a liquidus temperature of 1775°C so that it is seldom tapped from the furnace. The next lower region, 7, is the high-carbon ferromanganese alloy product that is poured through the metal tap hole. The lower side region, 8, is mostly carbonaceous material that probably accumulates from the furnace walls.

The role of the MnO-melt layer can be postulated only from observations of furnace behavior during operation. With excess carbon, the MnO layer is denuded progressively until the previously steady MnO content in the slag would suddenly decrease and cause considerable resistance problems as the carbon built up. This, in turn, would cause tapping difficulties. The converse is also true in that, with insufficient carbon, the MnO layer increases to a climax as the coke is depleted. This high MnO content raises the liquidus temperature of the slab, which again leads to tapping problems. In other words, since the MnO melt appears to provide a buffer region, the MnO in the slag is insensitive to the carbon balance until a critical point is reached. Radioisotope tracer tests using ^{54}Mn are being carried out at Metalloys to study the function of this MnO-melt region.

Independent mineralogical investigation of numerous samples suggests that the predominant reduction mechanism is via solid carbonaceous particles that are present in the slab and in the MnO-melt layer. Some under saturation of carbon and very low silicon values in the ferromanganese may be due to the metal carbide descending through this MnO-melt, which would partially refine the metal carbide.

Chemical analyses proved to be of limited value in view of the heterogeneous nature of the entire sample. In some photographs of metallographic specimens small granules of MnO of a characteristic green showed clearly as they were nucleated out of larger grains with a mostly iron slag. At the metal/melt interface the Mn-Fe ratio would be around 7.3%, rather high. The marked segregation of the melt, as determined in this investigation, indicates some lack of validity to the usual assumptions of homogeneity in the process. The furnace had cooled over a period of two weeks so that the pattern of segregation that could take place in this large-scale cooling process may contribute to the observed heterogeneity. In answer to a question as to whether the investigative group believed that the MnO layer existed during operation, Barcza replied that they had observed some MnO granules coming out of the tap hole with the melt.

In general he felt that the study provided an improved understanding of the smelting behavior of the raw materials, which has led to better control of the carbon balance and furnace resistance and, hence, to stable furnace operation at higher loads.

Ferrosilicon Smelting

Ernest Breton [72], a consultant to the Foote Mineral Co. of Exton, Pennsylvania, has described some of the intricacies underlying current controversies about proper operation of electric ferrosilicon submerged arc smelting furnaces. There is a wide divergence of opinion among furnace operators and designers and also a total lack of correlation among the existing records. Direct measurement of process information such as temperature, materials flow and yield of product normally used for feedback control and analysis are not available. Residence times of reactants in the furnace are variable, running from several hours to days. Because gas-phase conduction of electricity is involved, the electrical characteristics of the furnace fluctuate widely, often within seconds. This has made it extremely difficult to determine cause and effect relationships needed to design a control system.

To gain a clearer understanding of the interrelationships of the many factors that enter into the process, the Research Department of the Foote Mineral Company studied the operation of a 57-MVA ferrosilicon furnace at its Graham, West Virginia plant. The furnace was extensively instrumented during a 30-day run of 75% ferrosilicon with the following observables monitored continuously: furnace power, VARS and phase volts; electrode current, volts and harmonics of volts; stack temperatures, carbon dioxide in the stack and gas velocities. All recorders were synchronized with a time pulse so that data could be put into the same time frame for analysis. Later, a data logger was constructed and installed to continuously display and record individual electrode power, resistance, phase angle, current and volts.

Attempted correlations among changes in tap volts, electrode holder position and carbon theory with furnace performance were fruitless. Clear correlations only surfaced from a statistical analysis of the data. This was done with a computer using data keyed at half hour intervals, correlated in three ways: (1) correlations of metallurgical data, (2) correlations of electrical data, and (3) development of a physiochemical model to account for observed relationship.

The most significant metallurgical correlation was that between furnace power and rate of reaction. This was calculated from tons of charge consumed per hour instead of using tons of alloy tapped because of the long residence times of metal within the furnace and the time it took to weigh tapped metal. Daily production would be lower because of losses. Figure 5-3a shows the rate of ferrosilicon production (as calculated from input) and the corresponding energy consumption as a function of furnace power. There was an optimum power level beyond which energy consumption per ton increased sharply. This effect is attributed to excessive temperatures that occur when the rate of adding electric energy exceeds the rate of transfer of reactants to the smelting zone.

Energy consumption turned out to be directly related to imbalance of power among the three electrodes, a common occurrence in ferrosilicon furnaces. This is shown in Figure 5-3b, where the difference between the electrode with highest power and the one with lowest power is plotted against energy consumption. At a 30-MW load, an imbalance of 10 MW was the result of one electrode running at 15 MW and one at 5 MW. According to calculations of electrode temperatures, the temperature of the outside surface of the 5-MW electrode in time would drop below 1540°C (2804°F). At this temperature, silicon monoxide within the smelting cavity would dissociate to slag (SiO_2) and silicon metal, with an increase in electrical resistance as a consequence. In this run of 75% ferrosilicon, minimum energy consumption occurred at the stoichiometric ratio of carbon to silica. An excess or deficiency of carbon increased energy consumption. To use electrical measurements to discern current flows within a furnace, the purely electrical behavior of the submerged electrode furnace was studied exhaustively.

The data and oscillograms suggest that current flows by several mechanisms. Power increased with tap volts up to a maximum and then decreased. A plot of phase voltage in neutral versus current showed positive and negative resistance realms. Under normal operation the voltage waveform was close to sinusoidal; however, a severely imbalanced furnace distorted the waveform. Such phenomena are best explained by the feature of plasma discharges, as described by Guile [73] when applied to arc smelting current flows.

Within the submerged electrode ferrosilicon furnace, current flow appears to be predominantly by glow discharge rather than the arcing commonly surmised, as shown by the voltage waveform. Arcing severely distorts the waveform, but glow discharge results in a waveform that closely resembles those of the 57-MVA furnace. These observations come from a laboratory simulation of the electrical circuit within a submerged electrode furnace. A window in the model provided a correlation of photographic observations of current flow with electrical measurements. Interestingly, the lowest power factors were associated with unstable conduction as current jumped around the electrode. Arcing would produce very low resistance through a small-diameter channel for high-current low voltage and low power. Glow discharge was more diffuse to produce a low-current high voltage and high power. The furnace characteristics were closest to those of the glow discharge. The model showed an arc with the electrode close to the melt, but as the electrode was lifted a transition to glow discharge took place.

Evidence of superarcs that occur at high-voltage gradients was apparent in

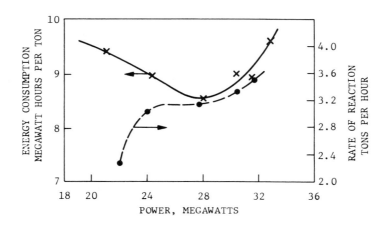

a) Relationship Between Power and Performance
of 57-MVA Furnace on 75% FeSi.

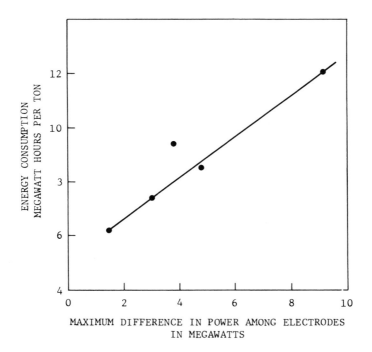

b) Effect of Imbalance of Electrode Power on
Energy Efficiency During a Run on 75% FeSi.

Figure 5-3. Ferrosilicon arc furnace energy relationships.

a study of electrode blows [74]. They are characterized by extremely large plasma plumes, which in the 57-MVA furnace could break through the surface as high velocity, white-hot blows. The vibration spectrum of one electrode was recorded under various operating conditions. During normal operation the electrode vibrated at 180 Hz. This was attributed to the interaction of magnetic fields of the three-phase current. When an intense blow developed, the vibration shifted from 180 Hz to 60 Hz.

The gas velocities and temperatures of the blows coupled with the 60-Hz vibration associated with them suggested that they were the result of superarcing in the upper part of the furnace. Hot expanding gases produced by superarcs would generate 60-Hz, 400-lb thrusts on the side of electrodes. Low-velocity, lower-temperature blows could be attributed to gases emitting from the smelting cavity.

It became apparent that chemical, metallurgical and electrical factors contributing to the smelting process were closely interconnected by temperature effects. Power input determines temperature. But temperature, in turn, affects power through its influence on the electrical resistance of the gas. Furthermore, temperature controls metallurgical reactions that subsequently affect electrical resistance. The operating electric current has a very large number of potential paths through the melt. The resistance must be controlled so that the current moves where it is desired and to assure the high dissipation glow discharge.

These complex interactions are depicted in Figure 5-4. In Figure 5-4a, interrelationships among the many factors that affect power are shown. All of the principal furnace controls affect resistance. Mix composition affects resistance of the burden, gas and the bath. Electrode position and tap volts determine the voltage gradient in the gas and, hence, whether conduction is by glow discharge or arcing. Figure 5-4b depicts the energy balance that determines temperature to show how temperature enters into the metallurgical factors of the smelting process. Temperature along with mix composition establish the configuration and location of the smelting cavity, thereby establishing the depth of burden. This affects critical temperature gradients within the burden needed for intermediate chemical reaction. Excessive temperatures lead to fusion of the burden and to high upper bed conductivity, with all of its deleterious consequences.

It has also been pointed out by Leonard Olds of Johns Mannville that high electrode erosion rates were a telltale indication of arc rather than glow discharge.

Production Models for High-Carbon Ferrochromium

Myles S. Rennie [75], of South Africa's National Institute for Metallurgy in Randburg, described the application of online data to computer control of a 48-MVA, high-carbon ferrochromium, closed-top furnace.

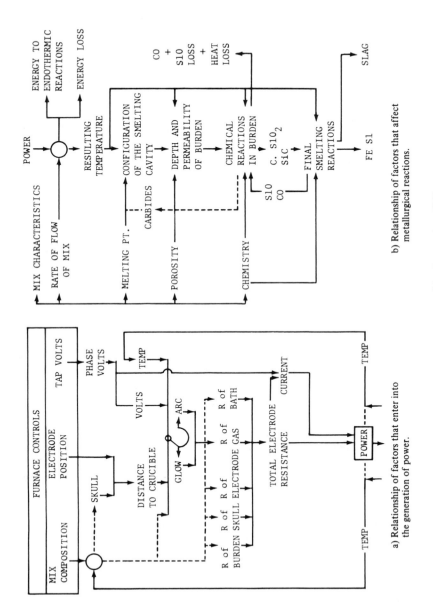

a) Relationship of factors that enter into the generation of power.

b) Relationship of factors that affect metallurgical reactions.

Figure 54. Ferrosilicon arc smelting interactions [72].

During the initial stages of the project, variations in the length of the electrodes led to unstable operation. This was due partly to imbalance between phases, which is particularly severe in large electric furnaces in which the reactance is greater than, or equal to, the resistance. It was very important to get an accurate assessment of the electrode arc region, outlined in Figure 5-5.

Models were developed to correlate electrode erosion with power and were updated to include raw material consumption and arcing. Although the agreement tended to be reasonable in most cases, there were still problems like point breaks, the variability of the electrode paste, and different rates of slipping and baking.

A simple model was developed that depended only on the geometry of the furnace and, using the resistance, calculated the height of the electrode tip from the metal bath. The agreement was found to be good under stable operating conditions but differed considerably under abnormal conditions. This led to a critical analysis of all the factors involved in the furnace operation.

Large variations in resistivity beneath the electrode tip were observed, both for short periods (hours) and for long periods (days). At the same time, considerable variations in the height of the metal bath were observed under each electrode; these variations were associated with the formation of individual craters and were related to the power developed under the electrode. To account for the variation in resistivity, small electrode perturbations were introduced. Subsequently, larger perturbations were used in the estimation of resistivity in various regions beneath the electrode.

The results showed that the small region just beneath the electrode tip was responsible for most of the voltage drop and, hence, power generation, and that high resistivity was due primarily to arcing. Small changes in electrode position produced large changes in resistance, as shown in Figure 5-6a.

This explained the sensitivity of the furnace to electrode control. Using the resistivity results, furnace operating characteristics could be derived from the modeling study, and the volumetric distribution of power appeared to be fundamental to the operation (Figure 5-6b). This was in agreement with the energy requirements of the metallurgical process. By use of the maximum current that can be carried by a Soderberg electrode of a given diameter as criterion, the results were applied to different sizes of furnaces. The agreement with operating furnaces was reasonable and showed that the power density was similar for different furnaces.

In answer to questions, Rennie added the following points:

1. As the electrode was lowered, voltage was dropped to avoid excessive currents.
2. It was difficult to sort out the various components of the current path, for example, between gas and slag. Due to the high throughput of feedstock,

there was not much slag in the crucible. Each arc creates its own melt, with feedstock insulating the electrodes from each other. Most of the voltage drop is in the gas. The coke melt layer would have very low resistance. As much as 30% of power might be dissipated in the plasma portion of the arc, with the remainder in slag and underlying material.

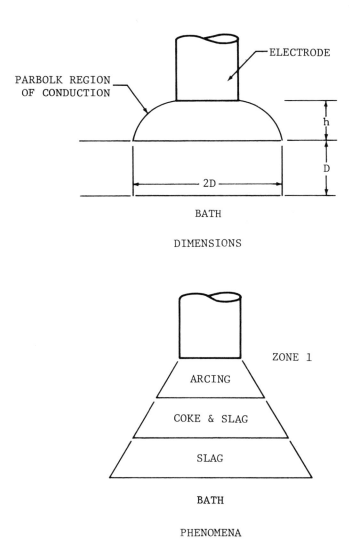

Figure 5-5. Electrode arc region in high-carbon ferrochromium production.

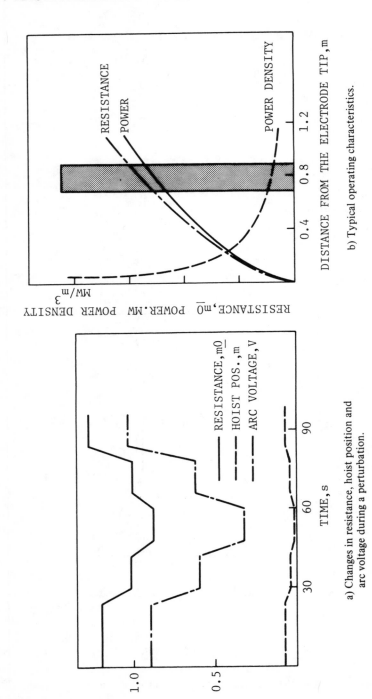

b) Typical operating characteristics.

a) Changes in resistance, hoist position and arc voltage during a perturbation.

Figure 5-6. Arc furnace electrical behavior [75].

MODELS OF ARC FURNACE OPERATION

To accurately understand and express coherently the nature of arc furnace behavior there have been many efforts to develop accurate mathematical models closely coordinated with the experiments. At the present state of development, the interactions between experiment and model are very similar to those of experiment and theory in areas of physical research where the phenomena are not yet clearly observed or fully understood in terms of what can be observed. The following recent examples of arc furnace model development illustrate to what degree the observations of arc furnace behavior are understood.

Mathematical Models for Soderberg Electrodes

While prevalent U.S. practice is to use electrodes of solid carbon or graphite for arc furnaces, in other countries frequent use is made of the more complicated Soderberg electrode, especially for high currents. In this electrode, carbonaceous material in paste form is fed continuously into a cylindrical sleeve, at the lower end of which it bakes into a solid to form the contact area for arc or glow electrical discharge into the melt. Initiating and maintaining such a set of circumstances through the entire smelting cycle necessarily requires very exacting knowledge of the physical and chemical state of the electrode. Leif Olsen and Reidar Innvaer [76] of the Elkem-Spigerverket R. & D. Center in Vagsbygd, Norway, described some material modeling tools that are useful in obtaining such knowledge.

The Soderberg electrodes can be quite massive, up to several meters in diameter, and temperature distributions can range from 80°C to 2000°C. Some examples are shown in the distributions for a stressed electrode in Figure 5-7. Direct measurements on operating electrodes is one approach. Temperature distribution up to a maximum of 1200°C has been measured in a number of electrodes. However, measurements are costly and difficult to carry out at the highest temperatures and during defined operating conditions. Mathematical models of the Soderberg electrode have therefore been developed and verified against measurements and data from electrodes in operation.

For temperature calculations, the electrode is divided into zones that run both radially and longitudinally. The zones are adjusted to the electrode geometry, material composition and boundary conditions. Each zone is subdivided to form a grid system of concentric ring elements. The main principle for the calculations is to compute the heat and current balance for each element. Boundary conditions for the electrode are input variables in

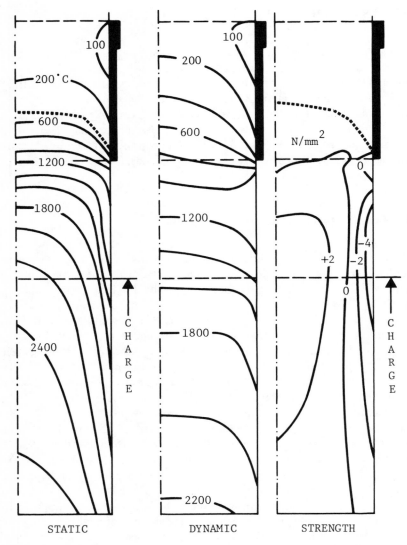

Figure 5-7. Calculated temperature (°C) and thermal stresses (N/mm², c tension, -compression) [76].

the program. Data for temperatures of the surroundings, heat transfer coefficients and electrical contact resistivities must be given for each case. Electrode current, slipping rate and properties of the electrode materials, such as electrical and thermal conductivities, are all important parts of the calculations. They are given as a function of temperature.

The program results may be printed as tables and plots of the following, all in spatial distribution:

- temperatures
- generated heat
- heat flux
- electric current
- electrical potentials

To get a better understanding of Soderberg electrodes in steady-state conditions, two computer models have been developed. The first one, ELKEM-S, is based on direct current, whereas alternating current is used in the second one, called ELKEM-X. Electrode parameters that can be varied in the calculations are as follows:

- current
- slipping rate
- electrode dimensions
- design of contact clamps and casing
- thermal boundary conditions, including cooling water temperature in contact clamps
- electric contact resistivity in holder area
- material properties

Figure 5-7 "static" shows calculated isotherms for a 1.55-meter-diameter electrode. Steady-state conditions in a ferrosilicon furnace are simulated. The models have been particularly useful in evaluating the position of the baking zone, i.e., 450–500°C, in the electrode. A reasonably high position of the baking zone is necessary to avoid breakages in the soft paste. Another application has been in determining the heat balance among furnace components, as shown in Table 5-2 for ferrosilicon and ferronickel smelting.

By contrast to the static models, a dynamic model, known as ELKEM-D, has been developed for unsteady-state electrode operations. ELKEM-X represents the start conditions for the dynamic run. In addition to the

Table 5-2. Heat Balance for Ferrosilicon and Ferronickel Furnaces
by the Static Model [76]

Heat (kW)	Fe-Si +	Fe-Si −	Fe-Ni +	Fe-Ni −
Generated	725		97	
Via Electrode Tip	20		13	
Via Electrode Surface		649		86
Consumed in Electrode		96		24
TOTAL	745	745	110	110

electrode parameters used in the static models, the following factors can be varied as a function of time:

- current
- slipping
- thermal boundary conditions
- electrode position in the furnace

The model has been used for calculating the cooling and reheating conditions in electrodes in connection with shutdown periods of the furnace. Figure 5-7 "dynamic" shows the isotherms 12 hours after steady-state operation ("static"). During this period of simulation, the electrode current was abruptly cut to zero for 6 hours, then restarted and increased during the next 6 hours to normal load. The dynamic model has been used primarily for studies of the baked electrode, but in the following cases the baking zone has been most important:

- small versus long slipping increments
- "forced slipping," i.e., high slipping rate at reduced current, and thereafter current gradually increased to normal load
- "long slipping," i.e., a long slip at reduced current, and current gradually increased to normal load

Calculations of temperatures can indicate only partially the thermal stresses in an electrode. A mathematical model for calculating the thermal stress distribution has been developed in which temperatures calculated by the dynamic model as well as the temperature-dependent mechanical properties of the electrode material are taken into account. The purpose is to simulate stop and restarting procedures to avoid hard electrode breakages. Again, a discrete element method is used for the calculations. The electrode is divided into concentric rings with triangular cross section. The elements are connected at common points, denoted "nodal points." Displacements of these points for each element can be expressed with matrices. Finally, the stresses in each element can be calculated.

The results are printed as tables and plots:

- stresses and strains in various directions
- displacements
- resultant stresses according to a fracture criterion
- stresses exceeding electrode strength

Figure 5-7 "strength" shows thermal stresses in the longitudinal direction of the electrode, based on temperature differences between "static" and "dynamic." The stress vector in this direction is in most cases dominating.

In one important example the behavior of two electrodes of different diameters was simulated through furnace cycles of 6 hours between start and restart. Higher stresses were formed in the smaller diameters. Material properties are very important with a low modulus of elasticity preferred to avoid breakage. As a whole, even though rather complicated, the mathematical models have been found to be very useful tools. In particular, they have been valuable in studies of the relative importance of the different electrode parameters on temperatures and thermal stresses. The trend toward larger electrodes with higher load will make the use of such models even more valuable.

In answer to a question on what material properties had the greatest influence on electrode failure, Olsen said that tensile stress and temperature dilation proved most important in the stress model. In the thermal model, electrical resistivity and its coefficient of change with temperature up to 3000°C were most important. In answer to another query, he said that the model could not treat differences in material microproperties such as particle size and paste composition. In a final comment he said that if a furnace has been shut down for 4–6 hours it should be brought on again to avoid irreparable deterioration of the electrodes.

Submerged Arc Furnace Electric Circuit Analysis

A. Bruce Stewart [77], of South Africa's National Institute for Metallurgy in Randburg, analyzed the electrical circuit of a submerged arc furnace to formulate a structure that eventually would include comprehensive, computer-implemented control. He emphasized the importance of furnace size in exacerbating the loss of balance among the interactions of the three supply phases. The measurement of the electrical variables of the load circuit, shown in Figure 5-8, is difficult, owing to the high currents flowing in the electrodes. This difficulty gives rise to errors in the measurement of electrode-to-bath voltages, which make it difficult to control the distribution of power within the furnace. Larger furnaces aggravate this problem: the parameters in an electrical circuit of large furnaces are more difficult to measure, and these furnaces are also more difficult to control.

To measure electrode-to-bath voltages on a submerged arc, leads are connected to the secondary terminals of the furnace transformer and to a copper electrode embedded in the furnace lining on the side of the furnace away from the tap holes. It has been shown that, if these leads are brought away from the furnace so that the loops formed by the measuring leads are symmetrical with the currents flowing in the electrodes of the furnace, errors in the measurement of electrode-to-bath voltages will be reduced. A measuring arrangement based on this theory has been developed and used on a large furnace to provide considerably more accurate measurements of the electrical variables.

In a submerged-arc operation, the reaction zone is completely buried so that all measurements of the nature of this region have to be made

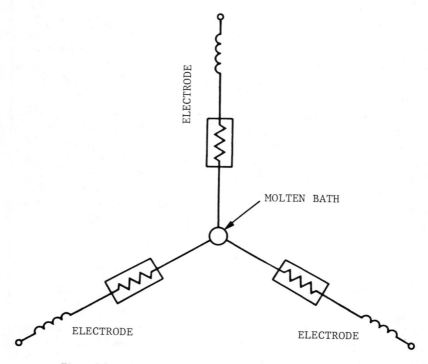

Figure 5-8. Equivalent load circuit for a submerged arc furnace.

indirectly. The accurate measurement of electrode-to-bath voltages provides a measure of the total resistances and generally inductive reactances of the load circuit only. Useful additional information on the nature of the arcs under each electrode can be obtained from the harmonic content of the voltage and current waveforms. Because there is an appreciable amount of interaction in the harmonics between the phases, measurements in one phase are affected by arcing changes in the other two phases. Nevertheless, this measurement has been used successfully for the analysis of various aspects of the load circuit.

The control of the power distribution in the furnace by resistance control or impedance control is unreliable, so that furnace operators usually resort to current control as the only means of operating the furnace. This method of control is satisfactory on smaller furnaces, where the relative reactances in the circuit are fairly low. However, on larger furnaces, where the relative reactances in the circuit are higher, there are large interactions in the power between the phases during unbalanced conditions, which makes even current control unreliable. This effect is shown and plotted in Figure 5-9a for seven different sizes of furnace. For each furnace, the relative power in each phase expressed as a percentage of the total power is computed for a current in phase 2, set at 80% of rated current, while the other two phases carry rated current. The increasing power imbalance with increasing furnace size is clearly evident. It should be noted also that the phase with the least current (phase 2) is not the phase with the least power (phase 1), which is a situation not often appreciated by furnace operators.

Arcing exhibits an interaction effect between the phases due to reactor imbalance. Changes in the arcing in one phase have an appreciable effect on the currents and voltages and, hence, powers in the other two phases. This effect is shown in Figure 5-9b, in which the powers in each phase are plotted for a variation of arc voltage in phase 1 from 0–100 volts, while the arc voltages in the other two phases are fixed at 50 volts. A change in arc voltage has a marked effect on the phase being changed (phase 1) and the preceding phase in the sequence of phase rotation (phase 3), while the following phase in the sequence (phase 2) shows little change. This effect is the same as that observed with no arcing, when a change in resistance in one phase affects the previous phase in the sequence of phase rotation.

Stewart emphasized, in conclusion, that the main problems are the high reactance of arc furnaces and the floating neutral condition of the melt itself. Measurement of the circuit conditions should be made at the electrodes and not at the secondary of transformers. In answer to a query, he hinted that electrode control is the way to utilize the analysis and measurement to obtain control. Arcing effects can be very destructive. They have observed arcs jumping to the side of the crucible on which the bath was lowered.

Electrical Considerations in Electric Arc Furnace Productivity

Continuing the emphasis on careful control of the electrical balance in an arc furnace, Gary M. Baker [78], of Keystone Steel and Wire in Peoria, Illinois, described the modeling work done for his plant. Baker contends that the imbalance of the three arc powers is a well-known fact and is due primarily to the unavoidable fact that the three phases cannot be constructed so that they are identical in impedance. His concept is to use computer simulations of the arc furnace circuit to theoretically equalize the power at the tip of each electrode in order to determine the correct settings for the electrode regulators in each phase.

The circuit of Figure 5-10 was used as model for the secondary side of the furnace transformer. Due to the use of a 125-Megavar synchronous condenser in this case, the buss can be considered, for all practical purposes, to be infinite at the primary of the furnace transformer. The impedances in the secondary side of the furnace transformer were obtained from nameplate data, and the furnace impedances (nodea 1, 3 and 5 to ground) were

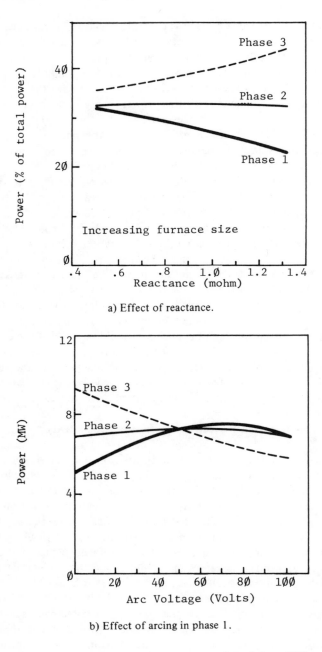

a) Effect of reactance.

b) Effect of arcing in phase 1.

Figure 5-9. Power imbalance in a submerged arc furnace [77].

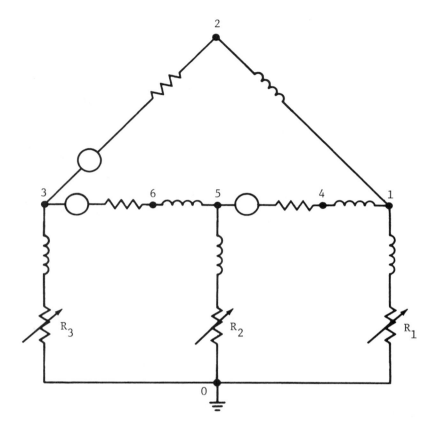

Figure 5-10. An arc circuit model [78].

determined from a three-phase dunk test. The resistances R1, R2 and R3 are variable and represent the furnace arcs in phases A, B and C, respectively.

By varying the arc resistances simultaneously from ten mohms to zero, power curves for each phase were generated as base information. The maximum power determined was 18 MW on phase B and 17.5 MW on each of the other two. Arc resistance at this point was 3.5 mohms. Because the Keystone shop operates under the so-called long arc-short arc method of operation, two points were investigated: one for meltdown and one for refining. The "meltdown" setting was to be 17 MW per phase, and the "refine" setting was to be 16 MW per phase.

Beginning with 3.5 mohms, arc resistances were varied independently and simultaneously until balance was reached at R1, R2 and R3. This point was 17.2 MW. Phase currents (transposed to primary values) were 1310, 1360 and 1312 amps, respectively. Phase-to-ground voltages were recorded for operating data (points 1, 3 and 5 to ground in Figure 5-10).

The same procedure was used for a "refining" setting. Arc resistances were varied with an aim of 16 MW at each arc. Balance was at 16.2 MW. Phase currents, again transformed to primary values, were 1550, 1600 and 1555 amps.

The model results were verified on the Keystone furnace. It is 6 meters in diameter and takes a 153-metric ton heat. Since the furnaces were somewhat overdesigned, there is little warping to cause concern. Verification of the model was done in actual operation by observing the arcs through the slag door on several heats during flat bath conditions. First, primary currents were set at 1350 amps each and secondary voltages recorded. The arcs were of different intensity visually. Hot spots were noted on a wall next to the B phase. The arcs were then adjusted to the theoretical values obtained in the simulation and again observed on several heats. Generally, the arcs were more in balance visually than previously, and the spots were not as apparent. Voltage checks matched closely with the computer simulations.

While phases one and three showed little difference in comparison with each other, B phase was noticeably "hotter" under normal conditions. By running B phase 50 amps ahead of A and C, arc powers are generally balanced under normal conditions. The B phase hot spot has, in fact, been reduced since this discovery, and adjustment of B phase regulator has proven to be a considerable refractory cost reduction over the same period the year before.

ARC FURNACE OPERATIONAL IMPROVEMENTS

Some of the recent improvements that have been effected in routine arc furnace operation are described in the following subsections.

Production of Silvery Pig Iron in Covered
Submerged Arc Furnaces

Walter T. Fairchild [79], of Foote Mineral's Ferroalloy Division, gave a very general description of his company's process for producing silvery pig iron with a pollution control system retrofitted to its arc furnace. He cautioned that he would not reveal various proprietary details of the engineering and design. The installation was first described in 1976. Now, after three years of operation, he could say that in the broadest terms it works, but not easily. To obtain continuous furnace operation, production rates had to be reduced by lowering the supply voltage.

In 1976 Foote Mineral Co. installed air pollution control equipment on the two existing open furnaces (15 and 20 MW) located at its plant in Keokuk, Iowa, and began producing silvery pig iron with the furnaces completely covered for the first time. The water-cooled covers installed on these furnaces have been provided with openings that accommodate three Soderberg electrodes, ten charge shafts, the "clean" offgas stack, an emergency dirty gas stack, and miscellaneous inspection and explosion ports. Gaseous, particulate laden by products of the smelting reactions are continuously drawn by fans through the "clean" gas stack into the venturi scrubbing system where the particulate material is separated from the gas stream. The cool, clean offgas is flared off, while the dirty underflow containing particulate material and condensables is collected in tar traps and pumped to a single recirculating water treatment plant. Here the particulate material is separated from the water by flocculation, sedimentation and filtration. The products of this separation are a dewatered sludge, which is discharged, and the clarified water, which is cooled and recirculated to the scrubbing systems. Makeup water is taken from the Mississippi River and blowdown water is returned to the river after additional chemical treatment. The pollution control process is outlined in Figure 5-11.

During the startup period on each furnace, it was found that as secondary voltage and kW load were increased, a point was reached beyond which the inability to properly control electrode position and penetration caused overheating of the offgas and resulted in excessive maintenance delays. To obtain reasonably continuous furnace operation with adequate control of offgas parameters, it was found expedient to operate at reduced power using transformer voltage taps in the lower delta range, when possible, or in the high wye range, when necessary. The availability and load for each furnace gradually approached, but never attained, that which had been experienced in former open furnace operations. This is shown in Table 5-3.

Scrubbing and Water Treatment

The furnaces are presently operated in such a way as to obtain the highest kilowatt load commensurate with continuous operation and adequate control of the gas scrubbing process. Pressure, temperature and composition of the offgas must be kept within specified limits to maximize scrubbing efficiency and minimize operating delays. Most of the major maintenance requirements are in the furnace cover, raw material charge shafts, the lower section of the gas offtake and the venturi fans.

The quality of the recirculating water used in venturi scrubbing must be carefully controlled in temperature, pH, suspended solids and dissolved solids. Chemical requirements for flocculation and sedimentation of the particulate material in the dirty water vary considerably because of

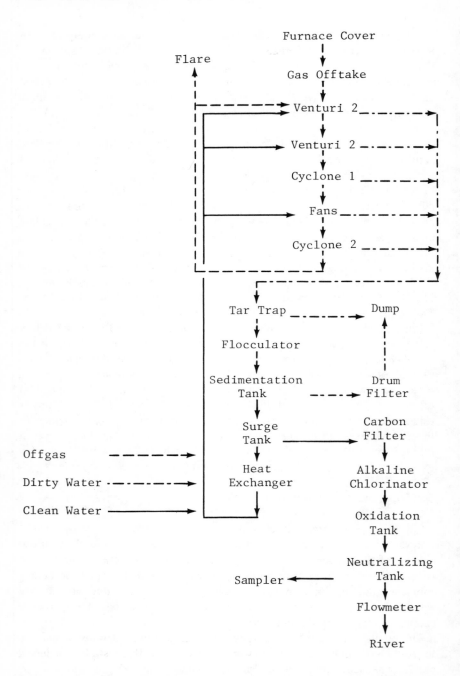

Figure 5-11. Pollution control process for silvery pig iron submerged arc smelting [79].

Table 5-3. Silvery Pig Iron Arc Smelting Operation Under Open and Covered
Conditions (kW load, availability, production rates–all grades) [79]

Furnace	Date	Mode	kW Load	% Availability	NT Operating Day
9	1975	Open	15,512	96.42	177
9	1976	Covered	13,001	76.35	146
9	1977	Covered	13,647	87.65	161
9	1978	Covered	13,653	91.83	160
9	1979	Covered	13,677	84.44	167
10	1975	Open	20,889	93.35	215
10	1976	Covered	14,265	89.06	139
10	1977	Covered	15,334	88.33	155
10	1978	Covered	17,298	92.40	182
10	1979	Covered	18,247	90.41	185

fluctuations in furnace load and raw material ratios used for each grade.
Blowdown water is discharged from the recirculating system, as required,
to remove either excess inbound water or water that is too high in undesirable dissolved solids content. As the water is discharged, it is further treated
to remove trace impurities that might harm the ecology of the river.

Ilmenite Reduction by a Carbon Injection Technique

Ralph H. Nafziger [80], at the Twin Cities Research Center of the U.S.
Bureau of Mines, reported on studies aimed at the efficient use of titaniferous
magnetite (Fe_3O_4) and ilmenite ($FeTiO_3$) as secondary resources for titanium
oxide. Since other reduction reactions usually have proved difficult, Nafziger
has proposed and is carrying out initial tests using a carbon arc with a carbonaceous reductant. There are three objectives to the research:

1. carbon injection to reduce more efficiently,
2. alternate carbonaceous reductants, and
3. minimization of adverse environmental influence.

Melts were made in a one-ton laboratory furnace, with tests run on six
reductants and three slab fluidizers. The furnace operates a single-phase
ac electric arc powered by a welding transformer. The ilmenite is fed in
continuously through a long graphite injector about 4 inches in diameter,
with a 1-inch orifice. Powdered ilmenite is poured in first to provide a
molten pool at the base of the crucible. Then the other injectants for
reduction tests are added. The basicity during the process ranges from 1.5
to 1.6. Metal and slab samples were taken during the test. In the pellet
charges, the reductant was mixed with the pellets.

The effect of reductants for preliminary laboratory-scale tests is shown

Table 5-4. Effect of Reductant in Reducing Ilmenite (percentages) [80]

	Reductant				
	Charcoal	Coal	Coke Breeze	Metallic Coke	Petroleum Coke
Yield	78	82	90	82	72
Slag/Metal	3.8	3.2	2.9	3.4	4.2
Fe_T	7.7	5.1	3.7	7.1	12
TiO_2	49	51	52	47	49

in Table 5-4. The best results were for the three coke reductants. Amounts of carbon are not very influential, except for some iron reduction at high levels. A similar test was carried out on the effect of conditioners. The use of fluorspar (CaF_2) with yield at 90% gave overall best results. A mixture of CaO and oxides yielded 82%. Alkalis yielded 74% and no conditioner at all resulted in 86%.

Moving from these preliminary results, larger-scale tests were then carried out for coke breeze, coal and coke as reductants. The materials were inter-mixed in the crucible according to four distinct techniques:

- rabbling (manual stirring of the melt while reductant is added)
- pelletizing
- injection with CaF_2 conditioner
- injection with no conditioner

Results are shown in Table 5-5 for yield, energy usage and carbon utilization. Energy expenditures are large due to the small furnace. Nafziger concluded from the results that carbon injection did offer some promise, especially when taking into account a reduction in process effluent that does not show in the table. Since the work is still preliminary, the economic promise aimed at in use of a secondary resource cannot yet be assessed. The technical reasons are strong and indicate that full-scale processing would benefit by the technique. During injection, the reductant was introduced directly into the metal, which promoted a greater rate of carbon dissolution and homogenization, owing to a more effective mixing action. In addition, the reductant did not directly contact the slag and no "insulating" screen of residual ashes was present on top of the slab bath during injection to affect carbon dissolution. Therefore, better control of carbon dissolution can be realized when carbon injection techniques are used.

Melting and Continuous Casting Operations at Armco

C. L. Cundiff, Lawrence L. Ludwig and Robert B. Webb [81] reported on Armco Steel's operations in Kansas City. The broad issues of productivity were addressed in the context of a declining market for steel in the

Table 5-5. Large-Scale Tests [80]

Reductant	Material Handling Techniques			
			Injection	
	Rabbling	Pelletizing	CaF$_2$ Conditioner	No Conditioner
Coke Breeze				
Yield (%)	65	27	73	64
Energy (kWh/ton/LM)	1916	1651	1990	1370
Carbon Utilization (%)	59	24	73	62
Coal				
Yield (%)	69	70	62	42
Energy (kWh/ton/LM)	1686	1736	1650	1700
Carbon Utilization (%)	65	64	60	34
Coke				
Yield (%)	56	56	60	38
Energy (kWh/ton/LM)	1511	1762	1535	1698
Carbon Utilization (%)	59	53	62	37

last decade. Armco's plant was a deliberate effort aimed at increasing productivity in its cast steel operations. Its Kansas City operations date back to 1886 and now comprise the largest steel mill in the U.S. west of the Mississippi. Products generally are parts for agricultural, mining and construction equipment, railroad spikes, nails, wire, steel floor joists, etc. Webb claims more finished products than any other steel industry plant in the U.S.

The electric furnace operations are the culmination of a project started in 1974 in Armco's Kansas City Works No. 2 Melt Shop. The two furnaces were placed in an open area flanked by a scrap yard. The scrap is conveyed through the shop. Basket levels and scrap hopper levels are coordinated for easy transfer. Pouring ladles can rotate 180° in 45 seconds. There is an elaborate transfer control pulpit with finished bars conveyed along the side, rather than the normal practice of under the pulpit. Webb cited these and other details as all-important for productivity. He also emphasized personnel practices, such as advanced training for operator, supervisory and maintenance personnel.

The finished plant was designed for 450,000 metric tons/yr. It was finished on time, with a small budget overrun. Yields, as a percentage of design goals, are 93.3 for the turret mill, 97 for the continuous caster and 95.4 for the furnace. The quality is high throughout the wide variety of products. All parts of the plant run 96% of the time. The next anticipated improvements are water-cooled panels and magnetic stirring.

New Class of Packaged Vacuum Arc Melting Furnaces

The LECTROMELT Corporation [82] of Pittsburgh, Pennsylvania has available consumable electrode vacuum ore furnaces, which have been designed for economic continuous large-scale production. They operate automatically and require only a 20-minute loading time. Furnaces with melt capacities of less than 50 pounds to more than 40 tons are available as production units.

All operating functions, including electrode and ingot handling, are performed at a single level. Manual operations are a minimal part of the cycle and the electrode control system requires minimum attention during melting. The highest degree of deoxidation and purification can be obtained without being limited by secondary reactions such as occur, for example, with crucible melts. Ingots of essentially the same composition as the finished product serve as an electrode to a water-cooled copper crucible. The electrode progressively melts and forms an ingot in the copper crucible, automatically controlled to keep the process at optimum conditions. As the metal is melted from the electrode, entrapped gases such as hydrogen and nitrogen are removed by the vacuum system. A carbon-oxygen reaction results in the chemical reduction and removal of oxide inclusions. Non-metallic inclusions tend to float in a molten pool at the top of the ingot. In addition, the ingot solidifies progressively from the crucible bottom to minimize localized segregation with the result of more homogeneous structure and composition.

COMPUTER-CONTROLLED ARC FURNACE OPERATION

The most important class of arc furnace operation improvements are those related to automating their operation. This is the special capability that distinguishes arc furnaces when compared to blast furnaces, which offer major economic advantage at a scale 10–20 times larger. There is no other category of electric furnace developments which elicits such a large, intense audience response at U.S. technical meetings. Some recent presentations are summarized in the following subsections.

**Operation of a Two-Furnace Ferrosilicon Plant
Under Process Computer Control**

C. T. Ray and W. L. Wilbern [83] of Union Carbide have reported on successful process control of a two-furnace ferrosilicon plant. They described the results of four years of operations under closed-loop process computer control of a 50% ferrosilicon plant complex, consisting of two 50-MW, submerged-arc electric furnaces and their auxiliaries at the Ashtabula, Ohio plant of Union Carbide Corporation, Metals Division.

An entire ferrosilicon alloys manufacturing operation at the Ashtabula, Ohio plant of the Metals Division of Union Carbide Corporation is being operated under closed-loop process computer control. Furnace No. 20, a 60-MVA covered unit, and Furnace No. 23, an open, hooded furnace with a 75-MVA power rating, are the two smelting furnaces in this production complex. Both furnaces are round, three-phase units with self-baking electrodes in conventional, triangular arrangement, and with fixed-electrode holder position. Furnace No. 20 has 55-inch-diameter electrodes and normally uses 45–50 MW in operation. Furnace No. 23 is equipped with 60-inch-diameter electrodes and is normally operated at a power input level of 50–55 MW. The computer also controls the operation of the automated mix batching and delivery system, the CO gas collection system of Furnace No. 20, and total plant power demand. Notably, the metallurgical requirements of the two furnaces, which make a variety of ferrosilicon products, have been programmed successfully for completely closed-loop computer control. Descriptions of the computer installation, the software package, and the functions of the computer have been reported by Wilbern [84]. Ray and Wilbern [83] have presented operating results on the Ashtabula plant.

In this installation, a single-process control minicomputer has been implemented to completely control the operation of two 50- to 55-MW ferrosilicon furnaces and their associated auxiliary subsystems. The closed-loop control extends from the metallurgical calculations of the mix batch composition, and mix batch weighing and delivery, to the final analysis of the alloy as it is tapped from the furnace. The control software programming technique used was to separate the various subfunctions of the total furnace operations into program modules, control each of these functions as a single identity, and pass the necessary information between program modules through data files and global, common-data tables. Therefore, the total furnace plant operation is divided into the following, separate process modules:

- Raw Material Batching and Delivery
- Metallurgical Calculations and Corrections
- Electrode Consumption and Slipping
- Electrical Power Input and Regulation
- Tapping Interval
- Offgas and Fume Collection
- Alarm Condition Monitoring and Correction
- Data Monitoring and Logging

The raw material batching and delivery systems for the two functions consist of 34 bulk storage bins, apron conveyors, vibrator feeders, hydraulic load-cell weigh modules, conveyor belts and batch delivery cars. Each furnace has a series of surge bins located above the furnace, which have a capacity of 8–12 tons of weighed mix. Radioisotope level sensors monitor the level of mix in these bins and signal the computer when more mix is required.

Once per minute the computer scans the level sensors on each furnace. When mix is required for one or more furnaces simultaneously, the computer determines which bin to give priority based on the rate of usage from each bin, determines the raw material bin and quantity of each raw material, and transmits the batch request to a programmable controller. The programmable controller verifies the batch request and weighs the raw materials simultaneously into the weigh modules. The controller transmits the actual weight of each material back to the computer, which logs the data and performs a weight comparison to provide correction for weighing error. Should a raw material bin run empty, the computer will direct the controller to weigh from another bin containing the same raw material, or from a bin that has a permissible substitute. After the weighing is complete, the raw materials are delivered to a common hopper, where the total weight is compared to the sum of the individual weights. Once the check weight is performed, the batch is delivered to the furnace.

Metallurgical control is based primarily on raw material analysis, product analysis and electrical characteristics of the operating furnace. Each tap from the furnace is sampled as the metal is being tapped. The sample is sent to the plant laboratory for analysis, which is then relayed directly back to the furnace computer from the X-ray spectrometer. Taking trends into account, the furnace computer compares actual to desired analysis and adjusts the batch composition of the raw materials to the furnace if the final product does not meet specifications. Raw materials are also adjusted if the electrical characteristics indicate an impending poor furnace condition or unbalanced operation.

Neither furnace at Ashtabula has electrode head position regulation. Elimination of the mechanical, electrical and hydraulic equipment, and the flexible power conductors necessary for electrode holder position regulation, results in significant initial capital cost savings and reduced maintenance costs. However, without the ability to move the electrode holders, control of electrode slipping must be much more precise. The tip position of the electrodes relative to the hearth is controlled on the basis of observed electrical characteristics and a calculated, self-refining consumption rate factor. The computer slips the electrode as frequently as 1 inch every 45 minutes to maintain a desired penetration in the furnace. Baking current is monitored to ensure that the electrode is not overslipped, since this could result in a paste leak or electrode breakage.

Maximizing and balancing power input to the furnaces is performed through online, voltage-changing transformers and by metallurgical adjustments. Utilizing the ability of the computer to predict what will happen when the voltage is changed on a transformer, based on historical data, the computer regulates the power input to a maximum without exceeding the electrical limits of the current-carrying components or the electrodes. Metallurgical adjustments are imposed if maximum load or balance cannot be obtained through voltage regulation.

At Ashtabula, the ferrosilicon furnace crucibles are tapped based on the energy input since the tap hole was last plugged. Using optical pyrometers, the computer "senses" whether the furnace is being tapped. A light warning system, operated by the computer, signals tapping personnel when a tap is impending and when to open the tap hole. The computer logs the tapping data when it sees the tap hole opened and closed. Automatic load reduction is imposed if the tap hole is not opened within limits, and eventual shutdown will occur if the limits are radically exceeded.

The offgas collection system on the semicovered Furnace No. 20 is controlled to optimize undercover suction, gas cleaning through venturi scrubbers, and utilization of CO as a by-product fuel in another process in the plant. The baghouse dust collection system, which collects fume from the secondary hood and tap hole of Furnace No. 20 and from the offgas hood of Furnace No. 23, which is an open furnace, is monitored for proper operation.

All critical components on the furnaces are monitored by either analog or digital inputs. Any analog input can be flagged for alarm action on either high or low limits or a rate of change, and any digital input can be set up to trigger alarm action on either the open or closed status. An alarm condition or failure of any critical component results in either an immediate power reduction on the furnace or a shutdown of the furnace. The alarm restriction is not removed by the computer until the alarm condition is cleared.

Numerous data logs are available on the furnaces. Every analog input in the system is scanned on either 5- or 60-second intervals. Values are averaged over various intervals up to 24 hours. Instantaneous values can be displayed on operator demand. Critical furnace data, such as raw material usages, power input, tap analyses, electrode slip and operating time, are logged and printed every 8 hours, 24 hours, weekly and monthly.

Operational Results

The degree of improvement that can be attained through the application of computer control naturally is dependent on the quality and uniformity of operation from which one starts. The comparative results, shown in Table 5-6 for the two furnaces, reflect this difference. Prior to the addition of computer control, Furnace No. 20 had experienced several years of excellent performance on 50% ferrosilicon operation. On the other hand,

Table 5-6. Operating Results of Union Carbide Arc Furnace Computer Control [83]

Key Operating Parameter	Improvement Under Computer Control	
	No. 20 Furnace (%)	No. 23 Furnace (%)
Operating Time	3.3	3.3
Average Operating Load	6.0	8.3
Kilowatt-hours per Pound of Alloy	1.4	3.0
Pounds of Electrode per Pound of Alloy	2.0	22.3
In-grade Production	1.9	0.9
Labor Hours/Ton of Alloy	13.3	13.3
Tons of Alloy	10.6	14.1
Cost per Pound of Alloy	3.6	5.2

following its initial startup, operation of Furnace No. 23 had suffered through an extended period of mechanical equipment problems, electrode failures and operator learning curve requirements.

Through providing a more stable furnace operation, the computer has reduced production costs and increased productivity. A constant electrode penetration in the furnace, along with a highly successful computer recovery technique following furnace outages, has resulted in a greater operating load, more operating time, less power usage per pound of alloy, and less usage of electrode per pound of alloy. More frequent and consistent metallurgical analysis of the furnace and more accurate raw material weighing have resulted in a higher percentage of salable product. Maximizing power input and balance has resulted in more tons of alloy from the furnace. A major unquantified benefit, so far, has been the improved safety for the operators.

In addition, Ray and Wilbern [83] made the following points:

1. Total cost of the computer system is 20 man-years and about $1.2 million. About $450,000 is in program development, which need not be repeated.
2. The installation can operate about 99.5% of the time under computer control. The operators are very well pleased now, although in the beginning they were suspicious that the computer would replace them, rather than serve as a tool.
3. The computer system was built by Mod Comp of Pampano Beach, Florida.
4. Electrode length was estimated by a combination of voltages, currents and electrode tip positions.
5. At individual electrodes, the operator controls stoking for an individual mix consistent with the batch mix set by the computer.

Productivity of the Steelton Electric Furnace

Spencer Blevins [85], of Bethlehem Steel in Steelton, Pennsylvania, described the productivity improvements at his company's electric furnace. The issues the company has been dealing with have changed drastically in the last 20 years. Most recently, the emphasis has turned to computer-run automation of the operations, similar to, but not as comprehensive as, Union Carbide's system, described by Ray and Wilbern [83]. The computer is used primarily to provide exact and complete information to the operators. Their resultant control is very detailed, even to include house-keeping status. This has facilitated management critique and provided appropriate advice for improved operations. Since 1969 the plant has produced 26 million metric tons, with production rate rising from 38 to 49 metric ton/hr.

New Computer-Controlled, Ultrahigh-Power (UHP) Arc Furnace of Krupp Stahlwerke, Sudwestfalen AG, Siegen, Federal Republic of Germany

A comprehensive computer control operation, very similar to that of Ray and Wilbern [83], has been reported by G. Kroncke, Rudolf Baum, R. Heinke and S. Kohle of Krupp in West Germany [86]. A newly con-structed meltshop had to replace an old, open-hearth furnace shop with four 100-metric ton furnaces, with a capacity of 3500 metric tons per month. Main reasons for the replacement were viewpoints of environmental control and questions of economy. The aim had to be to beat the conversion costs of the open-hearth furnaces, which had no gas cleaning facilities, with only one dust-cleaned, UHP arc furnace. Furthermore, the new arc furnace meltshop had to fit into the infrastructure of the open-hearth furnace area, that is, between scrap yard and casting pit.

The tap weight was predetermined by the crane capacity of the existing casting pit with 105 tons of liquid steel. The active shell diameter is, therefore, 6.8 meters, including outside-installed, water-cooled panels. The roof is also water cooled, except for a dome area around electrodes. The pitch circle diameter is set to 1400 mm, regarding the electrode diameter of 610 mm. Cable and bus bar copper cross section is adapted to 80-kA phase current. The transformer with intermediate circle switch at 30 kV and a rating of 75 MVA is connected on the primary side to 110 kV. The highest secondary voltage is 750 V. The electrodes are 0.61 meter in diameter, with 2.6-mohm estimated circuit reactance. At cold spots of the furnace there are three natural gas burners installed for supplemental heat. Process control of the high-powered furnace started in 1976.

A computer controls the meltdown of the scrap charged by two buckets and the refining of the steel in the furnace, and collects all data for the next heat. All important process signals and data are recorded in the control room of the furnace. The most important measured and calculated values are indicated, depending on the procedure of the heat. The charge materials are predetermined for each heat, with minimum costs according to the chemical composition of the to-be-produced steel grade. These data are the basis for the calculation of the necessary oxygen and energy demand. Alloys to be added are calculated for the addition in the ladle. The precalculated data of alloy and flux additions are introduced automatically into the alloy feeding plant.

A thermal model of the charge materials in the furnace is calculated by the energy input and losses. The energy balance considers, besides the electrical energy input, the input through jet burners and lance oxygen. Also, exhaust gas and cooling water losses are calculated. The energy input into the furnace is dependent on the thermal state, considering also the wall temperature in the hot spots. After meltdown, the bath temperature is continuously calculated and adjusted according to results of temperature measurements. The computer controls the power demand of all furnaces of the meltshop, with regard for the thermal state of individual furnaces. Power demand control influences the furnace control and reduces the energy input into the furnaces by adjustment of the tap voltage, or by lifting electrodes from the melt.

In general, during the first 18 months of operation the results exceeded expectations, boosting the production rate from 42 to 65 metric ton/hr. It is anticipated that recent market conditions would allow full utilization of the furnace capacity to a rate of 41,000 metric tons monthly. These advantageous results can only be reached by dividing the operating procedure into one part in the furnace and another part in the ladle, coursing the furnace to be a machine for melting and dephosphorization, allowing additions and bath temperature homogenization. The rapid melting practice, in connection with the computer control, results in short tap-to-tap times of about 100 minutes and consumption figures for lining materials below 1 kg of bricks and 5 kg of gunning material per ton, and below 4 kg of electrodes per good ingot ton. Energy consumption is about 520 kWh/ton.

Further advantages of the computer control are an equalized melting and refining procedure, more complete oversight of all operational events due to the permanent recording, and possibilities of statistical evaluations. Heatlogs are written automatically, and alloy additions calculations are carried out for each determination of the chemical composition.

In answer to queries, the following points were added:

1. Furnace utilization, from the standpoint of its capability, appears to be around 97%. If the computer is down, power must be reduced 50%.
2. The level of jet burner application influences the arc length that must be used.
3. The quality of the continuous cast product is very high in order to meet requirements of the automobile industry, its main customer.
4. The "as cast" billets do not have to be treated or conditioned in any special way.
5. Jet burners are best to preheat scrap in furnace charge. Inside the furnace, uniform melting is obtained.

6.   Limits of the electrode, at present, are the barrier to using higher currents for a higher production rate.

PLASMA PROCESSES

Plasma discharge processes for heat treating melting and for smelting are technically elegant for the high-quality product they yield but still not generally economical enough to be applied in such large markets as production and treatment of steel. This section describes some of the most recent developments that have been discussed in the U.S.

Plasma Processing of Ferromanganese Slags

A group at the University of Toronto, Kwazi Donyina, J. D. Lavers, A. McLean and R. S. Segsworth [51], reported on their methods of utilizing the slags of the high-manganese slag process for producing high-carbon ferromanganese. They aimed at reducing the major impurities normally contained in the ferromanganese product: silicon, phosphorus and sulfur.

For the experimental study, a closed-hearth extended arc flash reactor (EAFR), patented by Alcock and Segsworth [87], was used. As shown in Figure 5-12, it consists of a 20-cm-diameter hearth formed within a high-grade magnesia refractory lining. Three hollow graphite electrodes at 120° spacing supply up to 30 kW of three-phase power to the furnace hearth. Crushed high-manganese slag, mixed with appropriate amounts of reducing agent (low-grade coal or other sources of carbon) and flux, is fed into the reactor through a rotary feeder, which also acts as a preheater. As the charge materials fall through the rising hot gas stream into the zone of diffuse plasma, considerable preheating and prereduction take place. Metal and slag from the reduction reactions collect in the hearth below the plasma zone and are heated by radiation and convection from the plasma arcs. In operations for the experiments it was found necessary to add argon to the plasma region to remove harmonics from the voltage and current waveforms.

Experimental results using the extended arc flash reactor are shown in Table 5-7 for the following tests:

a)   Fluxless high-acidic smelting of high-manganese slags. Special grade silico-manganese contains 14–20% Si and standard grade silicomanganese 20–25% Si. Plasma-grade silicomanganese produced here contains less silicon and more manganese.
b)   Smelting of high-manganese slags with flux addition (lime) to reduce acidity. Low-silicon ferromanganese is produced with similar carbon contents.
c)   Phosphorus removal. Several experiments were made in which phosphorus was deliberately added to the charge in the form of calcium phosphate (Heats 3-1 to 3-3) or phosphorus pentoxide (Heats 3-4 to 3-6). Even with

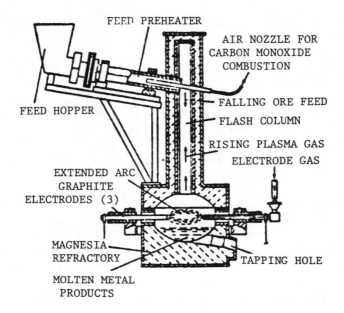

Figure 5-12.   Extended arc flash reactor [44].

phosphorus additions up to 8.3 pct, the alloys produced contained less than 0.005%.

d) Sulfur removal. Excess sulfur in the form of ferrous sulfide was added in several experiments. From the results shown, it is clear that significant desulfurization occurred in all cases.

There are indications that in addition to the foregoing impurity reduction, the plasma method described may also yield sizable energy savings in a full-scale unit. The energy required to produce 1 metric ton of alloy from slags in the 30-kW reactor was about 8 MWh. From additional experiments conducted in a 100-kW unit, the energy consumption was decreased to approximately 4.5 MWh. This may be compared with a value of 5 MWh/metric ton of silicomanganese produced from slag and ore in a conventional 30-MW submerged arc furnace. The main reasons for low residual impurities, efficient energy utilization and high recoveries of manganese, cited by Donyina [51], are as follows:

- improved burden preheating in the preheater and the flash column
- full utilization of the furnace cross section for the reduction process
- clear-cut separation of melting and smelting processes
- high slag temperatures and, thus, high lime slags
- fast reduction rates leading to high furnace productivity
- low manganese loss due to protective slag cover

Table 5-7. Experimental Results on High-Carbon Ferromanganese
with Extended Arc Flash Reactor [51]

a) Fe-Mn-Si Alloys

Heat #	Wt% Mn	Wt% Fe	Wt% Si	Wt% C	Slag[a] V-Ratio
1-1	65.81	19.00	6.59	4.2	0.8
1-2	72.48	13.40	10.98	3.1	0.7
1-3	76.92	11.58	8.59	2.9	0.8
1-4	70.32	7.33	8.13	4.5	0.8
1-5	77.98	6.42	8.74	2.9	0.7
1-6	73.99	9.96	8.92	2.8	0.8
1-7	78.92	5.58	6.36	3.4	0.9

b) Low-Silicon Ferromanganese

Heat #	Wt% Mn	Wt% Fe	Wt% C	Wt% Si	Slag V-Ratio
2-1	79.88	13.41	4.8	0.005	1.6
2-2	81.51	5.78	4.5	0.30	1.7
2-3	80.75	8.93	3.1	0.28	1.2
2-4	87.26	5.96	2.2	0.23	1.5

c) Phosphorus Experiments

Heat #	Wt% P in Charge	Metal Analysis				Slag V-Ratio
		Wt% Mn	Wt% Fe	Wt% Si	Wt% P	
3-1	2.07	76.33	10.69	4.77	<0.005	0.8
3-2	4.14	78.70	11.32	3.96	<0.005	0.8
3-3	8.29	59.36	30.45	1.64	<0.005	0.7
3-4	0.25	72.01	18.32	6.98	<0.005	0.5
3-5	0.50	79.15	10.51	0.81	<0.005	1.2
3-6	1.00	52.08	41.54	0.005	<0.005	1.5

d) Sulfur Experiments

Heat #	Wt% S in Charge	Metal Analysis				Slag V-Ratio
		Wt% Mn	Wt% Fe	Wt% Si	Wt% S	
4-1	0.86	78.92	11.51	7.91	<0.005	0.8
4-2	1.01	77.64	10.45	8.14	<0.005	0.7
4-3	1.16	82.14	9.13	0.005	<0.005	1.5
4-4	1.30	76.53	13.41	0.25	<0.005	1.2

[a] $\text{V-Ratio} = \dfrac{\text{Wt\%CaO} + \text{Wt\%MgO}}{\text{Wt\%SiO} + \text{Wt\%Al}_2\text{O}_3}$

There are also the following points of operational detail to be considered:

- Feedrate of material was kept at greater than 100 g/min. Argon was initially injected to stabilize the arc to a length of 8–10 cm. After steady conditions were achieved, offgas from the melt would provide a self-stabilization.
- The voltage and current waveforms are not affected by carbon in mix. Some work is continuing on their characteristics for a variety of particles allowed to fall through the system.
- With respect to possible contributions by varieties of hydrocarbons toward the reduction process, the experiments have not revealed any difference over a number of different types of coal that have been tried.

High-Rate Carburizing in Plasma Discharge

Recently, William L. Grube [88] of General Motors presented the latest developments in GM's program on the use of hydrocarbon plasma discharges to introduce very uniform carburized surface layers on parts with very complex shapes. In earlier work, Grube and Gay [89] had shown that carburizing at high temperature (1040°C) in a glow-discharge methane plasma simultaneously reduces carburizing time and natural gas requirements, while achieving uniform carburizing depths. Higher temperatures reduce the processing time. Required natural gas is reduced by the following combination:

- reduced process time
- electric heating of workpiece (induction heat cited as example)
- nonconventional carrier gas
- reduced pressure on surface being treated

The use of the discharge plasma has enabled them to reduce the gas pressure much farther than in simple vacuum carburizing, which offers the same advantages in reduced degree. Figure 5-13 illustrates the plasma carburizing circuit. The workpiece is the cathode of a dc discharge circuit. The anode is placed near the workpiece and, at pressures in the range of 1 to 20 mm Hg, a glow discharge plasma is established between them by application of a few hundred volts. The physical components of the glow discharge carburizing system are shown in Figure 5-14, as described by Grube [90]. These components are all similar to those of a vacuum carburizing furnace except for the electrode arrangement and the high voltage supply. A further essential addition is the control feedback loop from the temperature sensor to the glow discharge power supply. Since appreciable heating of the workpiece is due to electrical dissipation in the discharge, temperature control may be achieved by regulating discharge current while maintaining stable furnace power (heater or induction coil).

Sufficient carbon to produce a 1.0-mm case on conventional carburizing steels can be introduced in 10 minutes at methane pressures in the range 10 to 20 mm Hg. To reduce carbon concentration at the surface to an

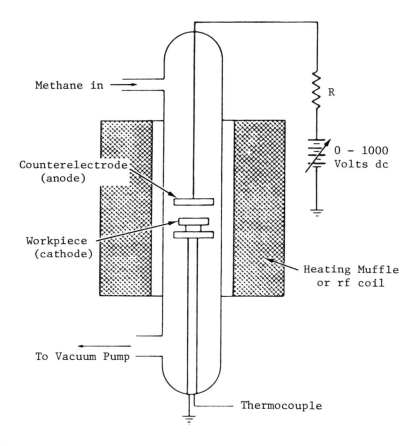

Figure 5-13. Plasma carburizing circuit [89].

acceptable 1.0 wt%, plasma carburizing is followed by a short (approximately 30 minutes) diffusion step. A diffusion model of the process indicates that the exceptionally high carburizing rates observed arise from the rapid infusion of carbon into the surface of the plasma during carburizing. Exceptionally uniform cases on surfaces of irregular geometry are achieved as shown in Figure 5-15. In subsequent work, Grube [91] showed that similar results could be achieved with heavier homologs of methane such as propane.

In private communication, Grube [92] said that he has not yet been able to quantify the economic advantage. This would require appropriate tests for higher production rates using higher-capacity equipment.

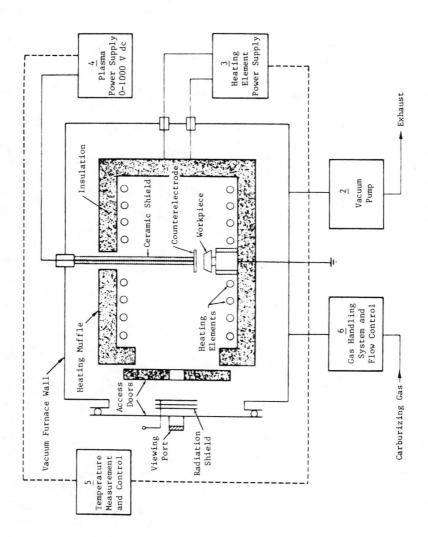

Figure 5-14. Glow discharge carburizing system [90].

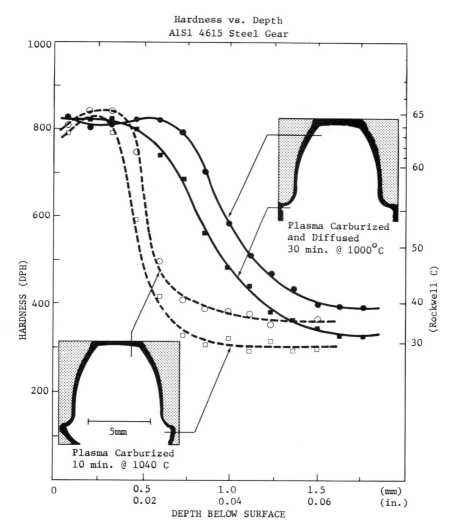

Figure 5-15. Hardness profiles through plasma carburized and plasma-carburized-and-diffused cases comparing tooth tip and fillet [89].

Plasma Smelting of Platinum

Texasgulf, Inc. has been testing a plasma smelting process for separating platinum group metals (PGM) from concentrates containing appreciable amounts of chromite [93]. Previous efforts to smelt PGM from chromite-laden concentrates failed due to chromite accretions on the furnace walls.

The South African state-owned National Institute of Metallurgy has been able to recover a PGM flotation concentrate having an acceptably low chromite content for a plasma smelting furnace process developed by Tetronic R&D of the U.K., known as Expanded Precessive Plasma. It is illustrated in Figure 5-16. The furnace contains a plasma gun, near the top, which rotates at 500–2000 rpm. The arc between the gun and the bath creates a plasma cone reaching temperatures of 20,000°C, causing near-instantaneous smelting reactions as top-fed concentrate enters the furnace. The fine-grained PGM concentrate gravitates through the plasma cone. Intense heat prevents chromite from adhering to the furnace, so that slag (containing the chromite) and metals collect at the bottom.

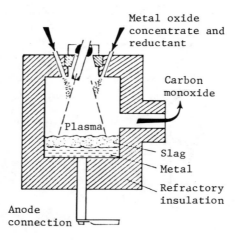

Figure 5-16.   Tetronic process for smelting platinum group metals from chromite-laden concentrates [93].

CHAPTER 6

RESISTANCE HEATING TECHNOLOGY

The use of ordinary direct current or standard power supply alternating current for heating by dissipation in conductive material has always been the simplest and least expensive of electric heating technologies. Hence, for small-scale manufacturing it has always had a natural market. It was inevitable that with increasing fossil fuel costs and curbs on industrial natural gas use since 1976 there would be increasing interest in larger-scale manufacturing and an added incentive for whatever developments would improve the technology.

There are three categories of resistance heating with distinct advantages and types of applications. The largest is the use of resistance heating elements in a simple furnace to heat the workpiece by a combination of convection and radiation in a chamber that is open to circulation. New developments here lie principally in the design and constituency of the heating elements, new insulation material and better integrated production arrangements. The most sophisticated category comprises vacuum or controlled atmosphere furnaces, which have always been in demand for special qualities in high-technology products. Developments here are centered on attaining higher temperatures and creating arrangements that accommodate higher production rates. The last category is the use of electric conduction in the workpiece material itself, variously called direct conduction or direct resistance heating. Here the new developments are coupled closely to the specific application, since the workpiece must be its own heating element.

The economic advantages favoring all types of elective dissipatory heating over the still intrinsically cheaper fossil fuel direct combustion lie in its control, high labor costs and current regulations for protecting the environment and maintaining occupational health and safety. These are not factors that can be treated generally or even generally within each of the three categories, since their quantitative impact is almost always keyed to the specific application. The following three subsections discuss the present technical advantages of each of the types of electric dissipatory heating and illustrate them with brief descriptions of some of the most recent and promising specific applications.

ELECTRIC RESISTANCE HEATING ELEMENTS

The new developments in resistance heating element use reflect mainly enhanced capability and higher furnace reliability, which lead eventually to reduced capital costs. For example, a series of very high-temperature open element atmosphere furnaces has been introduced by CM Manufacturing and Machine Tool of Bloomfield, New Jersey. Their CM 400 series can process a wide variety of metals and ceramics in all-hydrogen or hydrogen-nitrogen reducing atmosphere for processing metals such as tungsten, molybdenum, niobium, rhenium, nickel, palladium, zirconium and nuclear fuels. The 400A furnace operates to 1800°C and uses exposed refractory metal electric heating elements contained within a pure alumina refractory brick hearth. The 400Z furnace operates to 2200°C and uses exposed tungsten rod electric heating elements within a pure zirconium refractory brick hearth. Davidson [94] of Sylvania, has described the requirements for tungsten heating elements that operate up to 3000°C. Tungsten wire drawn from heated pure tungsten powder would sag and eventually part when heated to the range of 1300°C to 3000°C. This was due primarily to high-temperature grain growth that recrystallizes the long fibrous grains that are formed in the tungsten wire drawing process and that collectively bestow a macroscopic rigidity on the wire. However, it was discovered at Sylvania in 1963 that addition of minute quantities of potassium, silica and aluminum to tungsten oxide prior to its reduction to powder for drawing feedstock would result in greatly retarded grain growth on subsequent heating. In addition, Sylvania later began fabricating tungsten heating elements as a mesh of interlocking wire coils. The process includes 130 separate steps to control the tungsten wire crystal structure and provide maximum strength.

Lower-temperature elements have been improved by the development of a "porcupine" wire configuration at the Kanthal Corp. in Bethel, Connecticut [95]. Electric resistance alloy wire is folded and bent into a coil shape, which allows a generous airspace among the folds and bends. This has prevented excessive wire heating that curtailed the life of a heat gun in a prolonged operation. The elements are made of Kanthal A and A-1 alloy, with maximum wire operating temperatures of 1330°C and 1375°C, well above the nominal element operating temperature of 1000°C. Such elements also are used in electric resistance immersion heaters, which now can be used to heat a surprising variety of materials, as indicated in Table 6-1 [96]. To further the capability of immersion heaters, the M. P. Acee Co., Inc. of Natrona, Pennsylvania has applied an aluminum oxide coating to protect stainless-steel heaters

Table 6-1. Typical Immersion Heater Applications [96]

Materials Being Heated	Required Operating Temperature (°F)
Acid solution or electroplating tanks	180
Alkali & selected oakite cleaning solution	212
Asphalt binder, tar, other viscous compounds	200 300
Caustic soda 2% 10%	210 210
Dowtherm A flowing at 1 ft/sec or more Nonflowing	750 750
Ethylene glycol	300
± Fuel Oils Grades 1 & 2 (distillate) Grades 4 & 5 (residual) Grade 6 & Bunker C (residual)	200 200 160
Gasoline, kerosene	300
Liquid ammonia plating baths	50
± Lubrication oils SAE 10, 90-20, 120-30, 185, 100 SSU @ 130°F SAE 40, -50, 80, 80 SSU @ 210°F	250 250
Mineral oil	200 400
Molasses	100
Molten salt bath	800 to 950
Molten tin	600
Oil draw bath	600
Sodium cyanide	140
Heat transfer oils flowing at 1 ft/sec or more	500 600 600 750
Trichlorethylene	150
Vapor degreasing solutions	275
Vegetable oil (fry kettle)	400
Water (process)	212
Water (washroom)	140

even in environments such as molten zinc at 1250°F (676°C) [97]. The coating is applied as a plasma spray on a bond-coated surface that is thoroughly cleaned by a chemicals and pressure grit blasting.

One of the most effective new developments has been the substitution of ceramic fiber insulation for firebrick. Vanderkaay [98] has reported on a number of successful conversions at the Jameson Corp. of Kendallville, Indiana. In an oil-quench furnace about 1 cm^3 in volume, ceramic fiber lining 23 cm thick was substituted for 30 cm of firebrick. The superior insulation of the ceramic fiber allowed the power input to be increased from 145 kW to 166 kW, the maximum operating temperature to be reset from 596°C to 927°C, with a reduction in furnace shell temperature from 90°C to 74°C. On another furnace, ceramic fiber modules with a hung rod heating element system were used to convert a much larger furnace. The hooks support the hair pin-type elements at the top loop and lower sides, separated from the ceramic fiber by ceramic spacers. The elements are thus allowed to grow vertically during heatup with a minimum of horizontal movement. The major benefits of ceramic fiber in furnaces have been summarized by Bauer [99] of Watlow in St. Louis as follows:

1. Lightweight construction—it is one-third to one-half weight of insulating firebrick, one-tenth to one-twentieth weight of dense fireclay firebrick refractory linings.
2. Low volumetric thermal conductivity—ceramic fiber two-thirds thickness of firebrick conducts heat through at the same rate.
3. Low heat storage for fast temperature change—due to lightweight equivalent, ceramic fiber furnace wall has only one-fourth the heat capacity of firebrick.
4. High thermal and mechanical shock resistance—there is essentially no effect of rapid temperature or mechanical changes due to flexibility of the fibrous material.
5. High-temperature stability—there are no thermally sensitive mortar joints and ceramic fiber undergoes no appreciable secular changes under prolonged heating.
6. Facility of construction and repair—the ceramic fiber layers can be established as continuous sheets. Any repair or modification is easy to form and blend into the host structure.

Heat Treatment

Significance of the vanished economic advantage of direct fossil fuel combustion is typically illustrated by Grumman's recent conversion of a large autoclave to electric resistance heating, described by Romano [100]. The autoclave is indirectly heated by circulation of a heat transfer fluid in a nitrogen-blanketed closed loop, the heat transfer fluid now being heated electrically. It is used for curing polyimides, metal-to-metal and honeycomb-to-metal bonds, and composites with glass, boron and graphite fibers. Workpieces are subjected to heat and pressure in a cycle that usually totals 4–6 hours with 1100 such cycles in a typical year. The autoclave tubes are now heated by a Hynes Electric Heating Co. system with twelve 24-foot lengths of 12-inch-diameter pipe arranged in three 4-high stacks, all connected in series to form one continuous flow path, as shown in Figure 6-1. Seven 24-foot tubular heating elements are positioned in each 24-foot straight length of the twelve 200-kW stages. These are 3.5-inch-diameter tubes enclosing resistance heating assemblies, each tube closed at one end. The open ends extend out and are welded to junction boxes within which the 480-volt assemblies are connected in wye circuit. This arrangement permits access for removal of any resistance element that may require service without draining or otherwise disturbing the fluid system, or deenergizing the elements that terminate in other junction boxes.

The Hynes heater costs $150,000. To this must be added an equal amount for switchgear and main line power bus to connect to the utility grid. In the

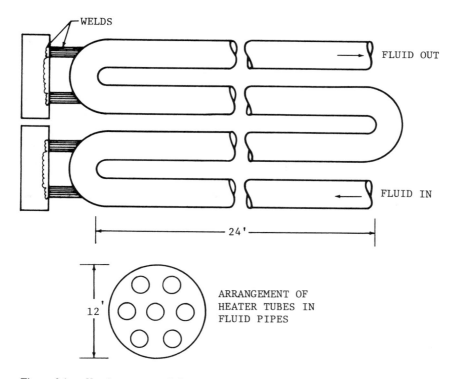

Figure 6-1. Heating system of Grumman autoclave after conversion to electric energy source [100].

gas-fixed system, the heating fluid was maintained at nearly 260°C and the system had to include a 4000-gallon accumulator tank, while a 1000-gallon expansion tank is all the electric heating system needs. System compactness is the principal reason why there are only 2500 gallons in the electric system. It is expected, incidentally, that useful life of the expensive heating fluid will be extended 20–25% by avoidance of even slight overheating, due to the accuracy of electric heat temperature control and uniformity of tubular element skin temperature with no hot spots.

Despite its 2400-kW size and the fact that it maintains autoclave temperature within 5°, the Hynes heater does not require an elaborate control system. Eleven of its twelve 200-kW stages have simple on-off controls actuated sequentially as demand rises and falls, with delay relays to avoid throwing full load on the transformer all at once. Only the first stage has SCR control, stepless from zero to full 200-kW output.

Recent developments in the heat treating of ductile iron have been summarized by Joe Ward [101] of Wagner Castings in Decatur, Illinois. In this area the heating process is technically independent of the energy source and its choice is indicated only by economic considerations. These, in turn, are driven at the present time by the availability of natural gas for industrial use. With deregulation of gas prices he expects it to cost as much as oil so that in a state like Illinois, where electricity is generated with coal combustion and nuclear energy, electric heating will begin to displace it.

George Otto [102] of Maytag in Newton, Iowa has described the use of electric resistance heating in sintering stainless steel. Maytag uses three furnaces for this application, two with graphite glass bar elements and a third with a graphite cloth element. These were very convenient and easy to maintain and repair while being capable of reducing the very high temperatures required for sintering stainless steel.

At the Delco Remy Division of General Motors, Jack Brewer [103] has been using electric resistance heating in carbon austempering of a clutch shell. The primary use of electricity here is in preheating. Delco Remy has bought no gas-fired equipment for eight years since it anticipates no availability of natural gas. In its operation, electricity is still more expensive than gas. Similar decisions in favor of electric resistance heat treating, despite its current higher overall cost, are being made at other firms. For example, an electrically heated annealing furnace to handle cold drawn steel at the rate of 578 kg/hr has been installed at the Handy & Harman Rathbone Co. plant in Palmer, Massachusetts [104]. The furance heats 2.28 metric tons to 1065°C and has provisions to cool it in a nitrogen atmosphere. It is computer controlled for heating and cooling rates, as well as time at stable temperatures. In a similar operation, the Natural Magnetics Corp. of Cerrator, California, uses a Honeywell DCP 7700 to control electric resistance thermal processing of alloy materials used in its transformer and inductor cores [105]. The alloys are heated in an atmosphere of hydrogen, which reacts with undesired inclusions. Cooling rates and other furnace parameters are tightly controlled to form grain and crystal structures, which enhance the alloy magnetic properties.

Hot Isostatic Pressure Treatment

Use of electric resistance heating elements allows for more complicated combinations of thermodynamic parameter variations. For example, the combination of heat and pressure (applied by gas on the outside of a casting), called hot isostatic pressing (HIP), was developed to eliminate waste in investment castings. It is an extremely significant development in metalcasing, as reviewed by Leonard [106]. Such castings can replace forgings in turbine components, for example. The temperatures used are high enough to homogenize the microstructure so that it appears similar to that of a wrought or forged piece. Both soundness and ductility are equivalent to forged materials. As pointed out by Freeman, of Howmet Turbine Components Corp., the forgings used in aerospace industries need a great deal of subsequent machining, which can be avoided in HIP investment castings made to much closer tolerances.

The development of equipment for HIP treatment at Industrial Materials Technology, Inc., in Woburn, Massachusetts, has been described by Widmer [107,108] and Price [109,110]. The pressure medium surrounds the workpiece heating element and interior heat shield, as shown in Figure 6-2. This allows the pressure-containment envelope to be maintained at ambient temperatures. The medium is usually an inert gas pressurized at 500–2000 atm. Internal metallic resistance heaters are most commonly molybdenum, michrome or Kanthal. A typical HIP sequence of operations would be as follows:

1. Castings or prepressed sintered aggregates are loaded onto fully instrumented support structures that must be carefully designed to avoid dislocation.
2. The assembly is placed inside the heating element, insulation, and outer envelope.
3. The cold system is closed, evacuated, flushed with argon and evacuated again.
4. The gas medium is then pumped by the compressor to some intermediate pressure so selected that subsequent heating to working temperature will cause a final increase in pressure to the desired working level.
5. Dwell time at peak pressure and temperature may vary between two and six hours, as illustrated in Figure 6-3. The power is then shut off to allow the system to cool slowly.

This company's experience to date is that almost any internal void space can be closed with the properly selected conditions of temperature, pressure and dwell time. If, however, a void space has a diameter close to the material

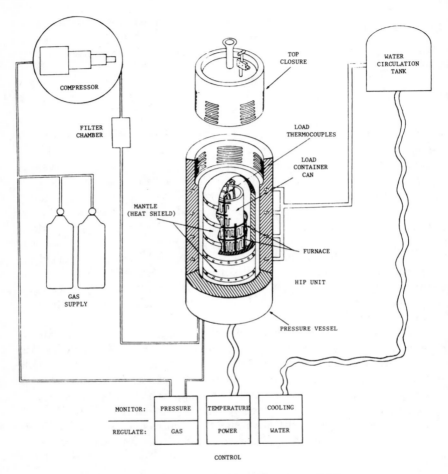

Figure 6-2. System for hot isostatic pressing [107].

thickness in a critical area, then the hole may migrate to the surface where it will not be influenced by the HIP process. The practice of titanium casting is growing at a very rapid rate. However, it appears to be virtually impossible to cast defect-free parts and, therefore, a routine HIP add-on process is being used for most titanium alloy castings in high-performance applications. Aluminum alloy castings face two kinds of problems: microshrinks and gas porosity, which is usually caused by hydrogen. Figure 6-4b shows the disappearance of the microshrinks under HIP. Hydrogen voids may also disappear under HIP, but can also reappear with a subsequent heat treatment under low pressure conditions.

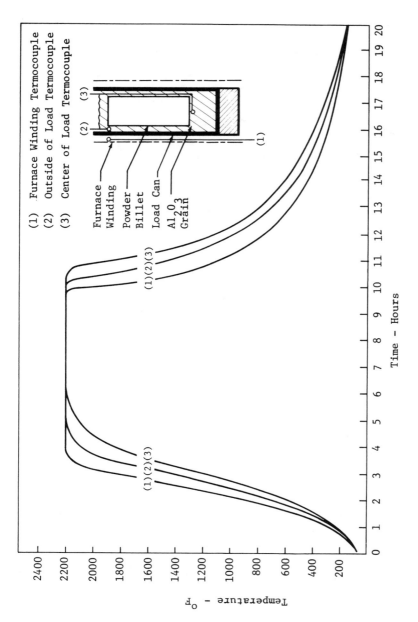

Figure 6-3. Instrumentation, heating and cooling of HIP furnace and load schematic [107].

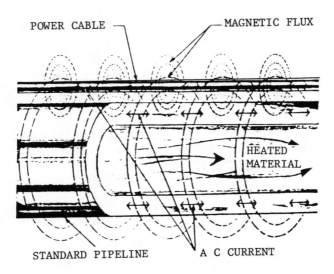

Figure 6-4. Electric impedance heating [112].

Among the current developments reviewed by Price [109,110] there is the completion of a very large pressure vessel about 2.44 meters high by 0.92 meters diameter. This was necessary for large-scale jobs that previously had to be turned away. In addition, they are trying faster-processing cycles and raising the temperature capability to 1815°C for refractory materials and ceramics.

Pipe Heat-Tracing

A fairly large use of electric resistance heating elements lies in the need to maintain flexibility in material being pumped outdoors over lengthy lines within a single large industrial plant. The fundamentals of meeting such requirements have been reviewed, for example, by Hutzel and Raychem [111]. The term "heat tracing" describes the selection, application, design, installation and operation of a heat source on a pipe or vessel. The fluid is heated to prevent freezing or maintain a necessary temperature. For example, water in drains or safety showers must be prevented from freezing. Caustic soda solutions must be maintained at temperatures to be effective in water treatment. Fuel ash, waxes and pigment oils must be maintained at temperatures that ensure low viscosity for pumping.

The standard technique has been to install a complete resistance element, adjacent to the pipe and surrounded by insulation. However, Koester [112] of Hynes Electric Heating in Kenilworth, New Jersey, has described a method of taking advantage of the pipe itself. He calls it electric impedence heating, the principle of which is illustrated in Figure 6-4. A small ac voltage is applied between the pipe and the single power cable. This cable can be outside the pipe's insulation because most of the electric dissipation is intended to take place within the pipe itself. This is possible for pipes made of relatively low-conductivity ferrous material so that heating is produced by magnetic hysteresis and eddy currents induced by the magnetic flux coupling between the outgoing and return current paths. Depending on the specific pipe metal, impedance heating will impart 20–35% more heat to the flowing substance than a resistance-heating pipe-trace system using the same amount of power. It has been applied successfully to pipes in a diameter range from 2.5 cm to 75 cm. It is most practical for pipe lengths of from 60 meters to about 5 kilometers.

In one of the most conventional applications, impedance heating may be used to heat pipes in a fuel oil unloading installation. The system includes 30-cm lines for barge unloading, storage tank filling and day tank fillings; a 15-cm fuel oil return line; a 25-cm forwarding pump suction line; and 40-cm and 50-cm transfer pump suction lines. All piping is insulated with 2-inch calcium sulfate under aluminum lagging. Because impedance heating does not need anything beneath the installation, preformed insulating block could be used on all liners. The system maintains fuel oil at a constant temperature of 60°C, with ambient temperatures down to 7°C.

VACUUM ELECTRIC FURNACES

The new developments in vacuum furnaces center on the attainment of higher temperatures and better control of the vacuum or special atmospheres. Durant [113] has provided a comprehensive review of heat transfer in vacuum furnaces. He emphasizes that radiation from the electrically heated element is the only thermal transfer and that one should aim at a cold-wall design with internal heating element to achieve both fast response and reduced heat loss. The furnace consists of work space surrounded in succession by radiant heating elements, which can vary in shape and material. Surrounding the work space and all heating elements would be a final unheated reflector, appropriate insulating materials and then the steel enclosure, which can be water cooled if necessary. The insulating materials can be ceramic fiber, which has no serious outgassing problem, as well as a low volumetric specific heat for fast thermal response.

The quick thermal response and low thermal inertia of the cold-wall vacuum furnace contribute to both productivity and energy conservation since loading of the furnace can be done at room temperature. Figure 6-5

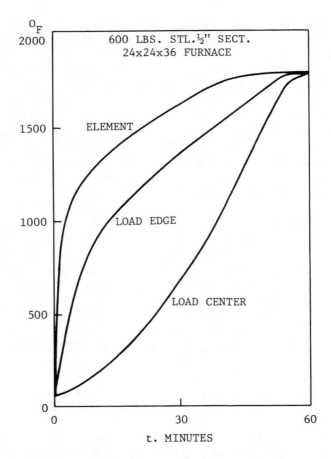

Figure 6-5. Typical response for heating steel in a vacuum furnace [113].

illustrates the rapidity with which a 272-kg mass of steel may be heated to 1000°C in a furnace of 0.34 m³ in volume. In the exemplary case illustrated, the energy actually delivered to heat the load during a sequence of heat runs represents 66.5% of the total energy used. Development of more efficient radiation shielding and insulation has sometimes improved the energy budget for vacuum furnaces by 15–20%. There are also arrangements of a number of chambers in vacuum heat processing system with isolation valves between loading vestibule, heating chamber and cooling chamber. These can also improve the heat budget by up to 20%.

Vacuum Brazing

A vacuum or environment for brazing operations provides the inherent advantages of the absence of oxygen and substantial outgassing of all materials involved in the process. There are several notable examples of recent developments that utilize these advantages.

Aluminum radiators have been under development since the 1920s without ever reaching large unit production due to inherent problems of joining in a high production rate process. The use of vacuum heating to eliminate the need for fluxes has been reviewed most recently by Warner [114] and Warner and Weltman [115] of ALCOA. Aluminum has an inherent oxide film that must be removed to join two pieces. For vacuum brazing, aluminum brazing sheets may be used clad on one or both sides with a brazing filler alloy. Some typical examples are illustrated in Figure 6-6, and other alloy systems for brazing are listed in Table 6-2. The four-digit numbers refer to basic aluminum stock types.

With oil diffusion pumps, the pressure in vacuum furnaces is reduced to the

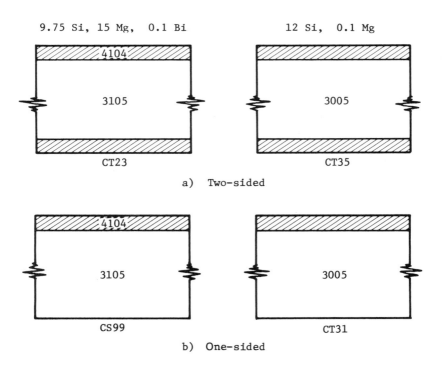

Figure 6-6. Clad aluminum brazing sheets [115].

Table 6-2. Alloy Systems for Vacuum Brazed Radiators [115]

Alloy	Sides Clad	Core	Si	Mg	Bi	Cladding Thickness Under 0.8 mm (0.032 in.)	Over 0.8 mm (0.032 in.)	Under 0.6 mm (0.024 in.)	0.6–1.6 mm (0.024 in.)	Over 1.6 mm (0.062 in.)	Fin Alloys
CT23	2	3105	9.75	1.5	0.1	15	7.5				3003,6951
CS99	1										
CT35	2	3005	12.0	0.1	–	15	7.5				5005,6951
CT31											
X14	2	6951	9.75	1.5	–			15	10	7.5	6951,3003
X13	1										
X8	2	3003	9.75	1.5	–			15	10	7.5	6951,3003
X7											
CS98	–	–	9.75	1.5	0.1			–	–	–	3003,6951

range 10^{-6} to 10^{-4} torr, with about half the gas as water vapor. As the brazing sheet is heated in the vacuum atmosphere the outer layers oxidizes, removing some of the oxygen from the system. When the sheet heats above 599°C (1038°F), ternary Al-Si-Mg liquid forms; some of this liquid exudes through the oxide. Magnesium vaporizes, gettering the system further. At about 580°C (1075°F) the amount of liquid in the cladding becomes great enough to flow. Some metal penetrates through the oxide and capillary action draws more liquid into the joint. The oxide skin in the fillet area is forced out, forming the fillet. The optimum brazing range for the sheets in Table 6-2 is 585-605°C. After this point the joint assembly may be cooled. The best joint properties are obtained by cooling from 480°C to 230°C (900°F to 450°F) in 15 seconds.

In one of the most advanced manufacturing technology areas, the vacuum electric furnace is used on an integral part of a highly automated procedure at Pratt & Whitney Aircraft of United Technologies in Hartford, Connecticut, for high-temperature turbine blades [62]. These blades run at 1372°C on a shaft turning at 10,000 rpm. The blades are cooled by pumping air through a hollow case. There is a great premium on trimming the blade to make it a better heat exchanger to air and so construct it to withstand stress and resist corrosion better. A great advance was achieved by splitting the blade in half, longitudinally. Casting it in two pieces gave much better dimensional control and the latitude to make thinner walls and trailing edges. The lighter blade also reduces root stresses where it joins the shaft and so it can turn faster at a higher temperature to raise engine efficiency.

A completely automated casting facility turns out the two-piece directionally grained turbine blades with virtually no human intervention. The blades are cast individually in a vacuum furnace in a process that is represented by 4000 parameters to be monitored and controlled insite entirely by a computer. An ingot of the turbine superalloy is melted by an electron beam gun and the melt drips into a water-cooled copper crucible from which it is immediately poured into the blade mold. The filled mold with a chill plate is transferred to a heater chamber within the vacuum furnace. Here the mold material solidifies from the bottom to the top by being lowered through the hot zone, creating a directionally solidified polycrystalline casting. The two blade halves are bonded together in a Pratt & Whitney development called the Transient Liquid Phase process. A boron-doped foil is applied to the blade halves by a photofabrication process and then the blade halves are clamped together using differential thermal expansion. They are heated in a vacuum electric furnace and the foil melts, diffusing the tiny boron atoms into the blade and creating a true solid-state weld without a heat-effected zone. These blades have at least twice the life of those formerly used. Moreover, there is possibly an even greater improvement that is anticipated. It is expected that the chill plate can be equipped to hold a seed crystal with a helical grain selector to assure only one orientation, as a single crystal throughout the mold. Lifetime improvements for such a blade would be an additional factor of five.

Vacuum Heat Treating

For heat treating, the advantages of vacuum electric furnace use lie primarily in the totally controlled environment coupled to greater energy efficiency [116,117]. There is no need for expensive auxiliaries such as hoods, ductwork and blowers. The combination of radiant electric heat, precision control of the heating element and the vacuum chamber environment allows faster processing. Finally, about 85% of the electric dissipation heat is delivered to the work load compared to 40% of the combustion energy in a gas-fired furnace. In addition, the system uses power only when processing; it requires no atmosphere conditioning, and weekend idling is eliminated. A pumped down vacuum furnace will remain conditioned for months. One of the most recent developments in these furnaces has been the availability of graphite heating elements in combination with a carburizing atmosphere for hardening purposes. The elements can operate at very high temperatures for rapid heating at high efficiency in a vacuum furnace. For a carburized gas, natural gas, nitrogen-diluted natural gas or nitrogen-diluted propane is back-filled into the chamber as a carbon source. With the minimum pressure to accomplish the work there is an excess carbon condition. At the end of carburizing, the furnace is reevacuated, allowing time and temperature for diffusion of carbon to establish case depth. Vacuum carburizing is relatively fast by two specific advantages:

1. The oxide-free surface condition of paste at the end of the soak period allows carbon absorption to begin immediately, without an oxide layer to break down.
2. During the diffusion cycle all carbon on the surface is diffused inward, with more lost to the atmosphere.

Some differences in actual carburizing rates between vacuum (Lindberg) and atmosphere equipment (Harris) are illustrated in Figure 6-7. Other recent examples of vacuum furnace advantages in heat treating follow.

Electric Vacuum Furnaces for Heat Treating Steel

To harden steel it is heated to its austenitizing temperature and then quenched or cooled rapidly. The peaking temperature will be between 815°C and 1280°C, depending on the grade of steel. Such temperature also must be controlled to within 14°C for commercial work. The heating must be scheduled so that scale formation on the steel surface is limited. In the final, most difficult step the steel must be cooled quickly enough to develop martensite, the hard phase of steel. Masters [118] has described how these

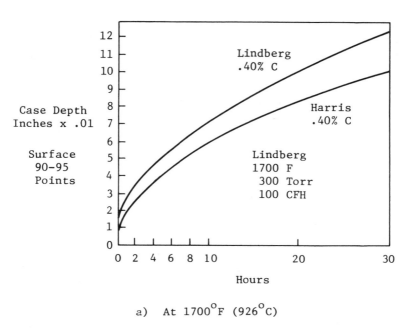

a)   At 1700°F (926°C)

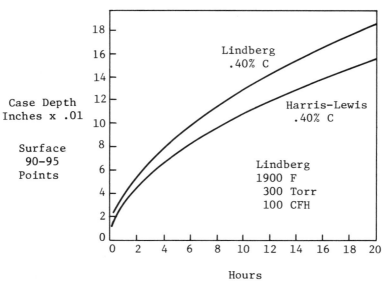

b)   At 1900°F (1038°C)

Figure 6-7.   Comparison of carbon diffusion rates in atmosphere and vacuum furnaces [117].

requirements are met economically using an electrically heated vacuum furnace.

When steel is heated to its austenitizing temperature as part of the hardening process, air must be excluded to prevent scale or decarburization. Either inert atmosphere or a vacuum will serve. The basic advantage of a vacuum furnace is its inherent simplicity. After heating in a very controlled manner, quenching is accomplished by turning off the vacuum pumps, backfilling with a cooling inert gas and using an internal fan, the only moving part, to flow the gas past internal heat exchangers and over the load, cooling it down to room temperature.

Programmed-Controlled Vacuum Furnace Annealing

To anneal transformer cores, the Siemens-Allis Power Breaker Division in Jackson, Michigan, is using a microprocessor-based digital control programmer with an Ipsen Model VFC vacuum electric furnace. A Honeywell DCP 7700 Digital Control Programmer enables preinserting up to nine different temperatures vs time programs for annealing. The combination provides the following advantages:

- control of the complex core-making process from core-winding through final annealing and transfer to core winding,
- control of delivery schedules, and
- tight quality control of annealing because the furnace vacuum prevents oxidation.

COMMERCIAL HEAT TREATING USES

Ross and Lobell [119] of Ironbound Heat Treating in Rosell, New Jersey, have reviewed a wide variety of applications for which vacuum furnaces provided sophisticated advantages at their plant. With the increasing volume of tool steels being heat treated, more stringent specifications in regard to surface chemistry, properties and physical dimensions of parts necessitated the installation of a vacuum heat treating facility to attain the quality required. There is a furnace for hardening and quenching under vacuums and two vacuum tempering furnaces, all designed and manufactured by C. I. Hayes of Cranston, Rhode Island. They use solid graphite elements cut in serpentine design. The outside envelope is water cooled. The principal advantages of the installation to the company are as follows:

- increased versatility for producing quality heat treated products,
- rapid heat treating cycles,
- heat treated parts that are clean and bright when leaving the furnace, and
- heat treated parts that are relatively free of distortion and held to a minimal dimension change.

Electronically Controlled High-Vacuum Furnaces

T-M Vacuum Products, Inc. of Riverton, New Jersey have introduced a line of high-vacuum furnaces that includes a solid-state electronic control system. Once programmed, the cycle is repeatable indefinitely and precisely until changed. Temperature is variable by stepless control to the limits of either 1000°C or 1300°C. The small size models are also available with a 500°C limit. Temperature uniformity over the hot zone is said to be well below 10°C, and a digital indicator on the panel displays the precise furnace temperature at all times. The units can be programmed for synchronized variations in vacuum, temperature and inert gas backfilling. Automatic start and stop features assure the proper sequence of starting and stopping steps. The furnaces are used in the heat treating, brazing or conditioning of a wide range of sensitive materials such as stainless steel, titanium, Ni-Span-C and super alloys.

Ceramic Fiber in Vacuum Heat Treating Furnaces

The qualities of ceramic fiber in low heat capacity, outgassing and high insulation have been utilized in a vacuum furnace for heat treating and brazing surface components at Alfa Romeo [120]. The furnace, built by Oxy Metal Industries of Great Britain, operates at 1400°C, 10^{-6} torr, with a volume of 91.4 cm drain by 91.4 cm high. The heating element is a graphite cage, and cooling is done by recirculating argon gas. A typical 180-kg charge is heated to 1200°C in 30 minutes and cooled in 35 minutes, with temperature controlled to ±2°C. Insulation consists of a 1-cm alumina fiber blanket backed by 2.5 cm of alumino-silicate blanket mounted between molybdenum sheet on the hot side and perforated steel plate on the cold face. The fiber blankets are Saffil, made by ICI America, Inc. of Wilmington, Delaware. At 1200°C there is a power saving of 10 kW/m^2, compared to a multilayer molybdenum screen construction used formerly.

DIRECT ELECTRIC CONDUCTION IN WORKLOAD

Of all three categories of elective dissipation heating, the last, utilizing electric dissipation in the workpiece material itself, intrinsically has the highest potential energy efficiency. Its advantages, both technical and economic, are very specific to each application since they depend on the material constituency, physical state and shape of the workpiece. The most prominent example is electric arc steel-making, already discussed in Chapter 5. The following few examples illustrate how its intrinsic advantages may be realized in other applications. Usually, the main technical barriers of the batch process nature and the interface problems of arranging an effective electrical contact to the workpiece must be overcome in an application that, like steel-working, places a high premium on efficiency of energy use.

Electroslag Remelting

Electroslag remelting, invented by R. K. Hopkins [121], is very similar to arc furnace steel-making. The process consists of melting the parent feedstock as a consumable electrode, with the melt collected in a shaped, water-cooled mold that yields a final casting close to net dimensions. It is illustrated, for a simple case, in Figure 6-8 [122]. The required heat for melting is produced by electrical dissipation in the electrode and melt usually with a single-phase ac supply at 30–50 volts. The use of direct current to remelt electroslag has been reviewed by Schlatter [123] of the Latrobe Steel Co. in Latrobe, Pennsylvania. Use of dc has no strong advantage over the use of ac, which is standard. However, it is useful when an ac furnace is not available. Typical electric energy requirements are 1200 kWh/ton, more than double the arc steel-making requirement. As the metal from the electrode falls through the

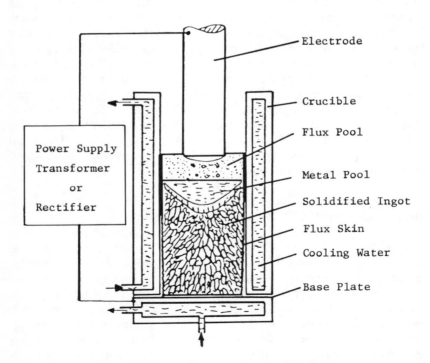

Figure 6-8. Electroslag remelting [122].

liquid slag it is refined with respect to sulfur, phosphorus and nonmetallic inclusive. It freezes from the liquid pool in the mold in a progressive, directional manner. This effectively eliminates macrosegregation; brings microsegregation to a minimum; and removes porosity, shrinkage, hot cracks and other defects. The principal disadvantage of the dc system is greater complexity of the electrochemical and thermochemical interactions, leading to chemistry control problems. Chemical homogeneity of dc-melted ingots is equivalent to that of ac-produced ingots, but the loss of easily oxidizable elements (Si, Mn, Al, Ti, Zr) depends on power mode, polarity and slag type. Experience with a variety of specialty steel grades has shown that significantly lower oxygen contents are achieved with the dc-rp mode. Table 6-3 is a comparison of ac and dc electroslag remelting.

Historically, steel castings were difficult to produce to desired shape and had poor mechanical properties due to segregation and porosity. Hence, forged steel parts were preferred. Recently, at small scales some newer casting techniques such as HIP overcame the old problems. It is in the large sizes (0.5–50 metric tons) that electroslag remelting is particularly valuable. The range of application for the process is very wide, depending, in small sizes, only on acceptable thermal gradients. In large sizes the limit is determined by those parts of the casting in which poor heat transfer prevents directional solidification. The apparent limits are that the cast steel sections must be greater than 5 cm but less than 1.5 meters and weigh more than 50 kg and less than 50 metric tons. An illustration of the properties of electroslag castings is shown in Figure 6-9 [124], a comparison with an air-melted bar in fatigue strength showing a marked increase for the electroslag remelter bar in the transferse direction. The improvement is attributed to better microhomogeneity and cleaner melt conditions [124].

Forging

Some of the advantages of direct resistance heating in a forge shop have been reviewed recently by Hill and Wilson [125] of IPE Cheston in Madison Heights, Michigan, who pointed out that there is almost no use of electric heating by radiation from furnace elements. Almost all forge shop electric heating is either by induction or direct electric resistance heating, in which the piece of steel to be heated is inserted physically into the secondary circuit of a large stepdown transformer. The IPE Cheston Plant uses only the latter method. The flow of current is along the longitudinal axis between points of contact. Steel has sufficiently high resistivity compared to the copper of the transformer to heat very rapidly. When the steel has been heated to an appropriate forging temperature, as determined by either a timer or optical pyrometer, it is disconnected and dropped onto a conveyor for delivery to the forge. Direct resistance heating has been found at IPE Cheston to offer the following specific advantages over induction heating:

Table 6-3. Comparison of dc vs ac Supply in Electroslag Remelting [123]

a) Sensitivities

Advantages of dc	Disadvantages of dc
Easier Cold Starting	Metallurgical Reactions Less Predictable
No Ingot Size Limitation	Chemistry Control More Difficult
Higher Melting Efficiency (sp)	Sensitive to Melting Parameters
Good Supply Line Balance	Prone to Electromagnetic Stirring
	Power Supply More Complex

b) Process and Product Characteristics

	Power Mode		
	dc-sp	dc-rp	ac
Cold Starting	Easy	Easier	More difficult
Slag Selection	Somewhat restricted	Somewhat restricted	No restriction
Ingot Size	No limitation	No limitation	Limited
Melting Efficiency	High	Low	Medium
Metallurgical Reactions	More complex, less predictable	More complex, less predictable	Predictable
Deoxidation	Poor	Good	Good
Desulfurization	None	Good	Best
Aluminum Pickup	Medium	Low	Low
Melting Variables	More restricted	More restricted	Wider latitude
Process Control	More important	More important	Less critical
Pool Stirring Tendency	Strong	Strong	Weak
Structure Control	Fair	Good	Good
Ingot Surface Quality	Good	Good	Good
Composition Control			
General	Satisfactory	Good	Good
Reactive elements	Poor	Marginal	Fair
Macrocleanliness	Good	Good	Good
Microcleanliness	Poor	Good	Best
General Hot Workability	Good	Better	Best
Mechanical Properties	Good	Good	Good
Consistency of Properties	Satisfactory	Good	Good

1. The energy consumption of a resistance heater averages 290 kWh/ton. When comparable 3-kHz or 20-kHz induction heating systems are utilized, the company cannot realize better than 350−450 kWh/ton of steel heated.
2. Because of its design simplicity and its ability to utilize 60-Hz power without conversion, the first cost for a resistance heating system is usually about 75% of a comparable induction heating system with similar automation. It should be noted that 3-kHz and 10-kHz induction applications, competing with properly applied resistance heating applications, are the most expensive types of induction heaters.

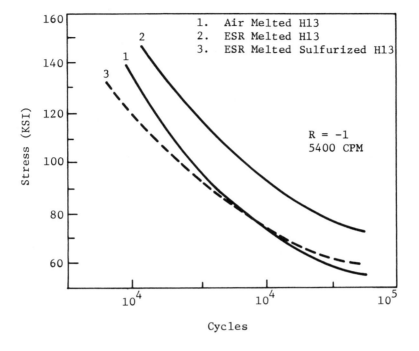

Figure 6-9. Fatigue strength curves for H13 tool steel from transverse specimens of billet [124].

3. Because of the high efficiencies of resistance heating, the water cooling requirements on a resistance heater are low enough that water recirculation and cooling are not required.

4. The modular concept that can easily be utilized in resistance heating offers a large degree of flexibility. Any one or all of the modules can be utilized at any time, even in a highly automated system. It is extremely difficult on most automated induction heating systems to change the number of modules being used if, for example, a malfunction should occur in one of the modules.

While induction heating can do anything that resistance heating can do, in some forging applications resistance heating is economically more attractive. These applications would have the following characteristics:

1. Diameters of stock are less than 3.8 cm.

2. The length to diameter ratios of the stock to be heated are greater than 6 to 1.

3. The total production requirements for heated steel are less than 2000 lb/hr.

If any of the above parameters are not met, the chances are that induction heating represents a better value in that application than resistance heating.

The fundamental reasons for this are that so many modules of resistance heaters are required to satisfy the production requirements that resistance heating becomes not only expensive but extremely difficult to set up and coordinate.

Internal Electrothermic Treatment of Steels and Alloys (IETT)

At the Pillar Corporation in Milwaukee, Wisconsin, M. Nankin [126] is developing a new form of heat treatment, utilizing direct resistance heating in a unique way. This method, unlike all other forms of heating, heats the entire body of the material homogeneously. Hence, the surface layer may be kept cool while the core is heated. In the typical application, IETT forms a new core structure retaining a case structure that had been formed by a preceding process. Depending on the initial core structure, it can be tempered, stress relieved, aged, solid-soluted or hardened. The treatment appears to have a unique ability to enhance simultaneously both fatigue strength and toughness. Moreover, it can be used with equal ease as nonferrous alloys or ferrous.

CHAPTER 7

MARKET OUTLOOK FOR ELECTRIC
HEATING EQUIPMENT

The variety of electric heating technologies in commercial use has clearly expanded since 1976, notably by the new importance for electron beams and lasers, and continued growth in the area of plasma testing. Within the established category of electric resistance heating, more refined methods such as vacuum brazing and hot isostatic pressure treatment are proving themselves surprisingly viable from both technical and economic viewpoints. In general, heat treating operations are becoming more varied and specialized as new advances in industrial technology increase requirements for different properties and characteristics of materials.

For heating equipment in general, and electric heating equipment in particular, there has been an increasing market within the general area of U.S. industrial technology. Some evidence for this can be seen in the preliminary data for industrial heating equipment recently released by the U.S. Department of Commerce Census of Manufacturers, 1977 [42]. The data are shown in Table 7-1, comparing the value of shipments in the years 1972 and 1977. To relieve the distortion in these data due to general inflation during the five-year interval, a third column is added showing what rise in shipment value would be expected based on the price rises characteristic of the overall sector of industrial commodities that includes heat treating equipment. The implicit price deflator over 1972-1977 for this sector, as calculated by the U.S. Department of Commerce Bureau of Economic Analysis, is 1.41 [127]. Heating equipment is included within the general sector of U.S. gross domestic investment expenditures, in the subcategory entitled Producers' Durable Equipment hierarchically related to the Gross National Product, as shown in Table 7-2a. With 1972 taken as a fiducial year, all price deflators in all categories are normalized so that they are equal at 100% in 1972 [127]. The deflator for Producers' Durable Equipment, as shown in Table 7-2b, has a slower growth than others and, therefore, indicates a declining share for

Table 7-1. Value of Shipments for Industrial Heating Equipment by all Producers
(1972 and 1977) [42]

Heater Type	Shipment Value ($ million)		
	1977	1972	1972+41%
Total	608.0	341.1	481.0
Electric Furnaces, Excluding Induction	136.6	67.5	95.1
Metal melting	21.0	19.0	26.8
Metal processing and heat treating	52.6	17.0	24.8
Infrared and extruding	60.6	28.5	40.1
Others	2.4	2.6	3.7
Fuel-Fired Furnaces, Metal Processing	131.0	97.2	137.0
Metal melting	24.0.	18.8	26.5
Metal processing and heat treating	54.8	48.0	68.1
Industrial ovens and furnaces for forging, etc.	51.0	35.2	49.6
Other	1.2	0.7	1.0
High-Frequency Induction and Dielectric Heaters	82.0	56.0	79.0
Radio-frequency induction, except metal melting	12.8	7.3	10.3
All other induction	62.2	37.1	52.2
Dielectric heaters	3.9	11.6	15.9
Others	3.1	0	0
Other Electric Equipment, Except Soldering Irons	172.0	91.0	128.2
Heating units and devices	97.7	47.8	67.3
Components of heating units and devices	72.1	37.7	53.1
Others	2.2	6.0	8.5
Industrial Furnaces and Ovens, Not Specified by Kind	86.4	29.4	41.4

capital investment in U.S. economic activity. Within this sector, however, as the data show for most categories in Table 7-1, heating equipment has a strongly rising share since the total shipments value has climbed 78%, compared to 41% for all Producers Durable Equipment in the interval 1972-1977.

Within the category of heating equipment only fuel-fired furnaces for metal processing fails to show the 41% expected rise. This is also true uniformly of all the fuel-fired heating equipment subcategories. Hence, the faster than average growth is clearly connected solely with electric heating equipment. In this category the smallest relative rise is shown by high-frequency induction and dielectric heaters, which indicates that electric resistance and conduction heaters, the other large categories, would have the most favorable market outlook. However, an exception should be noted among the subcategories. Noninduction metal melting furnaces are falling behind with only a 5% increase between 1972 and 1977, while induction metal melting advanced 68%

Table 7-2. Assumed Implicit Price Deflation for Heating Equipment

a) Organization of Categories

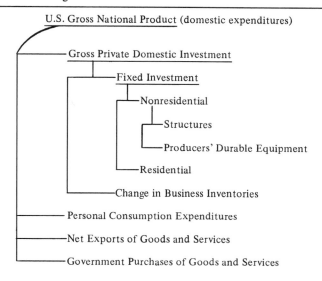

U.S. Gross National Product (domestic expenditures)

Gross Private Domestic Investment

Fixed Investment

Nonresidential

Structures

Producers' Durable Equipment

Residential

Change in Business Inventories

Personal Consumption Expenditures

Net Exports of Goods and Services

Government Purchases of Goods and Services

b) Implicit Price Deflator for Producers' Durable Equipment and Nonresidential Fixed Investment: 1971-1980

Year	Producers' Durable Equipment	Nonresidential
1980	168.2	179.9
1979	161.1	171.3
1978	150.3	157.8
1977	141.0	146.7
1976	133.3	138.5
1975	126.0	132.2
1974	109.2	115.3
1973	101.7	103.8
1972	100.0	100.0
1971	97.6	96.4

and radiofrequency induction advanced 76%. Dielectric heaters, with an actual decline of 66% are plainly responsible for most of the slower growth in the overall category of induction and dielectric heaters. This is due to the failure of dielectric heating to provide dependable economic benefit in its main areas of application: paper and fabric drying [128]. The subcategories with outstanding market growth are noninduction metal processing and the combined infrared and extruding. This last change is difficult to interpret because the infrared rarely is used for metal treating and must be separated from the extruding furnaces.

Another useful perspective is that of the industry that provides that treatment for metals as a service. Table 7-3 presents data on its billings over the years from two sources: the 1977 Census of Manufacturers and the Metal Treating Institute. Part a) of the table shows a 57% increase in the 5-year interval. The census data pertain to both commercial heat treating services and in-house departments, which would report to the census their interdivisional transactions. On the other hand, the Metal Treating Institute collects data on billings from those of its members who are all in the business of selling heat treatment as a service. These are shown in part b) of Table 7-3. There is clearly a much faster rise, with 113% gain over the 1972-1977 interval. In fact, if the commercial heat treating share of the total service value in part a) is subtracted, one finds that in-house heat treating services rose from $392.7 million to $471.0 million, for a rise of only 37.2%. The commercial heat treater's market share rose from 26% to 35% in the 1972-1977 period. Hence, a large part of the electric heating equipment buying rated above would be concentrated among the commercial heat treaters' plant expansion programs. However, the question remains why this shift has occurred from in-house to commercial heat treating companies.

Table 7-3. Metal Heat Treating Industry (Adapted from 1977 Census
of Manufacturers, U.S. Department of Commerce, 1979 [42]
and Metal Heat Treating Institute)

a) All Metal Heat Treatment Establishments: Value of Shipments
for 1977 Census of Manufacturers

	Value ($ millions)		
Product	1977	1977	% Increase
Heat treating, pickling, annealing, etc.	727.1	462.9	57.1

b) Billings by Commercial Heat Treaters for years 1973-1979

Year	Billings ($ millions)	% Increase from 1972
1972	120.2	0
1973	155.5	28.3
1974	193.2	61.0
1975	183.4	52.5
1976	207.7	72.8
1977	256.1	113.0
1978	311.5	159.0
1979	364.4	203.0

One must first recognize that higher electricity and fuel costs are eating up more operating funds of metal-working managers. For captive heat treating departments of large companies the decision is to refurbish the equipment or buy new equipment. For smaller firms the decision is not so clear cut because furnaces are a high-price item, with costs averaging between $200,000 and $500,000. As questions about the performance of in-house heat treating departments become more complex and the demand for stronger parts becomes more important to production managers, many smaller firms are asking whether it is worthwhile to continue the luxury of operating their own heat treating shops. They seriously question whether their operations are continuous and intense enough to fully utilize new capital equipment for heat treating. For example, a utilization of only 8 hours daily may not provide a high enough yield in cost savings to pay back the investment before the new gear becomes obsolete.

On the other hand, a commercial heat treater can arrange to keep his equipment operating continuously. Since heat treating is a very capital intensive industry (furnaces and furnace systems costing anywhere from $100,000 to $1,000,000 plus, depending on job requirements and type), many small firms do not feel they can or should make such investments to upgrade their operations. The problem now confronting the heat treating industry is the capability of existing commercial heat treaters to expand their services. Most commercial heat treaters are smaller firms with sales volume of between $500,000 and $2,000,000 and so are not geared to rapidly take over many additional customers. Every commercial heat treater that has been contacted was swamped with business: many had to refuse orders, and the word recession currently was not part of their active vocabulary [129].

From Table 7-3 it is clear that the commercial heat treating industry has had successive 20% growth years since 1976. The outlook is favorable for the future, although there are always questions as to the state of the economy and the various industries that support commercial heat treaters. A survey conducted in January 1980 indicated that commercial heat treaters had a very good year in 1979 and did not foresee any problems in 1980 [130]. However, there were some negative views on the future because of fear of recession, changes in government policies and rising energy costs. It seems that depending on the industry primarily served, a commercial heat treater will either be optimistic or pessimistic. Heat treaters in Detroit are having problems with one firm laying off 40% of its workforce, while other industries such as agricultural equipment, off-road vehicles, aerospace, mining and oil drilling equipment and tools are placing as much business as possible with their commercial heat treaters as they can handle.

A continuation of the trends in heating equipment indicated by the 1977 Census [42] in Table 7-1 is shown by the most recent data collected by the Industrial Heating Equipment Association. Table 7-4 presents the order reported by the members for the years 1974–1978. The industry as a whole is advancing at the same rate as in the 1972-1977 period. However, the direction may be turning a little away from electric equipment back to fuel fired. There are some further data released by the U.S. Department of Commerce that show the trend for shipments of industrial heating equipment continuing at least through 1979 [131], as shown in Table 7-5. While the 1969 total of $957 million can be compared to the 1977 total of $608 million in Table 7-1, it cannot be closely related to the limited sample order data of Table 7-4. The upward trend, according to Robert A. Ricciuti of the Commerce Department's Office of Producer Goods, was expected to continue in 1980 with estimated sales of $1.17 billion, for another 22% annual gain.

Table 7-4. Orders Reported by IHEA Members ($ × 1000)
(from Industrial Heating Equipment Association)

	1978	1977	1976	1975	1974
Fuel Fired	118,177	92,543	77,291	52,429	90,397
Electric Processing	71,368	67,960	35,824	43,562	23,765
Other Equipment	96,632	80,279	71,201	50,375	39,326
TOTAL	286,794	240,782	184,316	146,366	153,485
Electric, % of total	24.8	28.2	19.5	29.8	15.5
Fuel, % of total	41.2	38.4	41.8	35.8	58.8
Industry Growth	46	23.5	20.6	−4	1971-1974 8.8

Table 7-5. Profile of Industrial Heating Equipment Industry [42]

SIC Code: 3567 (furnaces and ovens)

Value of Shipments ($ millions)	957
Value Added ($ millions)	575
Total Employment (000)	17
Number of Establishments, Total (1977)	306
Number of Establishments with 20 Employees or More (1977)	124
Exports as a Percentage of Product Shipments	23
Imports as a Percentage of Apparent Consumption[a]	6

Compound Annual Rate of Change, 1976-1979

Value of Product Shipments[b]	10.2
Value of Exports[b]	8.0
Value of Imports[b]	14.9
Total Employment	1.2

[a]Imports divided by product shipments plus imports minus exports.
[b]Rate of change based on current dollars.

REFERENCES

1. Miller, G. "Coreless Induction Heating," *Electrotechnology Vol. 2, Applications In Manufacturing*, R. P. Ouellette, F. Ellerbusch and P. N. Cheremisinoff, Eds. (Ann Arbor, MI: Ann Arbor Science Publishers, Inc., 1978).
2. Barbier, M. "Electric Heat Treatment of Metals," in *Electrotechnology Vol. 2, Applications In Manufacturing,* R. P. Ouellette, F. Ellerbusch and P. N. Cheremisinoff, Eds. (Ann Arbor, MI: Ann Arbor Science Publishers, 1978).
3. Callaghan, D. D. "Utilization of Infrared Process Heat for Energy Conservation," paper presented at the 14th National Industrial Electric Heat Treatment Conference, Louisville, KY, March 3-5, 1980.
4. Szekely, J. "Role of Innovative Steelmaking Technologies," *Ind. Heating* 46(10):8-12 (1979).
5. Brondyke, K. J. "Energy Savings in Aluminum Production," *Ind. Heating* 29-32 (July 1979).
6. Gross, T. Personal Communication (May 14, 1980).
7. Laird, O. C. "Washington Update on Industrial Conservation," paper presented at the 5th ASM Heat Treatment Workshop, Detroit, MI, May 6-8, 1980.
8. Williams, J. Personal Communication (May 15, 1980).
9. Layton, R. E. "Technical Design, Application and Benefits of Automatic Iron Pouring Systems," paper presented at the 14th National Industrial Electric Heat Treatment Conference, Louisville, KY, March 3-5, 1980.
10. Stein, G. "General Motors Looks at the Future of Electric Process Heating," paper presented at the 14th National Industrial Electric Heat Treatment Conference, Louisville, KY, March 3-5, 1980.
11. Chandler, H. E., and D. F. Baxter, Jr. "Technology Forecast '79," *Metal Prog.* 28-71 (January 1979).
12. Gonser, T. Personal Communication (May 15, 1980).
13. Obrzut, J. J. "How Induction Heating Bypasses Energy Mess," *Iron Age* 55-58 (September 10, 1979).
14. Osborn, H. B. "Basics of Induction and Resistance Heating," paper presented at the 14th National Industrial Electric Heat Treatment Conference, Louisville, KY, March 3-5, 1980.
15. Sorensen, P. N. "Energy and Induction Heating," paper presented at the 12th IEEE-IAS Annual Meeting, Los Angeles, CA, October 2-6, 1977.
16. Bobart, G. F. "Electric Induction Heating for Metal-working," *Plant Eng.* 31(25):137-140 (December 8, 1977).

17. Balzer, N. Paper presented at the 5th ASM Heat Treatment Workshop, Detroit, MI, May 6-8, 1980.
18. U.S. Department of Commerce. "Preliminary Report, Annual Survey of Manufacturers–Fuel and Energy Use" (1977).
19. U.S. Department of Energy, Federal Energy Information Center. *Monthly Energy Rev.* (April 1980); *Energy Data Rep.* (January 1980).
20. Webley, K. G. "Economics of Induction Heating for the Forging Industry," paper presented at the 14th National Industrial Electric Heat Treatment Conference, Louisville, KY, March 3-5, 1980.
21. Jennings, R. E. "Forging Heat by Induction?" *Am. Machinist* 98-99 (November 1978).
22. Hutzel, J. "Automated Induction Machine Heats Various Size Billets for Forging Operations," *Ind. Heating* 46(9):36 (1979).
23. Duff, D. B. "Induction Heating for Automated Farming Operations," *Ind. Heating* 45(6):24-25 (1978).
24. Layton, R. E. Personal communication (May 15, 1980).
25. Schmidt, L. E. "Electric Induction Melting in the Gray Iron Foundry," paper presented at the 14th National Industrial Electric Heat Treatment Conference, Louisville, KY, March 3-5, 1980.
26. Finkelnburg, W. "Iron Foundry Installs Holding Furnace," *Foundry Management Technol.* 105:78 (November 1977).
27. Finkelnburg, W. "Induction Holding Furnace Improves Metal Quality," *Foundry Management Technol.* 107:80 (February 1979).
28. McKenna, J. P. "Application of Electric Furnaces for Aluminum Casting," paper presented at the 14th National Industrial Electric Heat Treatment Conference, Louisville, KY, March 3-5, 1980.
29. *McGraw-Hill Scientific and Technical Encyclopedia* (1978).
30. Pfaffman, G. "Precision Heat Treating with Induction," paper presented at the 14th National Industrial Electric Heat Treatment Conference, Louisville, KY, March 3-5, 1980.
31. Arnold, L. E. "Carburizing Heavy Section Mill Components," paper presented at the 5th ASM Heat Treatment Workshop, Detroit, MI, May 6-8, 1980.
32. Wolf, J. "Progressive Induction Hardening of Gears Submerged in Quenchant," *Ind. Heating* 46(8):10-12 (1979).
33. Hutzel, J. "Microprocessor-Based Programmable Controller Used in Thermal Processing of Alloys," *Ind. Heating* 47(3):38 (1980).
34. Hutzel, J. "Automated Induction Heat Treating of Ring Gears for High Production Rates," *Ind. Heating* 45(9):42 (1978).
35. Hutzel, J. "Design of Automatic System for Hardening Camshafts," *Ind. Heating* 45(1):24 (1978).
36. Craig, A. J. "Heat Treatment of Chain Saw and Small Engine Components," paper presented at the 5th ASM Heat Treatment Workshop, Detroit, MI, May 6-8, 1980.
37. Hutzel, J. "Three Induction-Melting Furnaces Will Provide Hot Metals for BOF at Breckenridge Works of Allegheny Ludlum," *Ind. Heating* 45(10):26 (1978).
38. Hutzel, J. "Superheating Hot Metal with Channel Induction Furnaces for QBOP and BOP Steelmaking," *Ind. Heating* 46(9):16-20 (1979).
39. Smith, R. L. "Induction Heat Treating and Equipment," paper presented at the SME Heat Treatment Conference, Los Angeles, CA, June 10-12, 1980.

40. Larson, C. "50 KHz Induction Heat Treating," paper presented at the 5th ASM Heat Treatment Workshop, Detroit, MI, May 6-8, 1980.
41. Conta, R. L. "A Process for Improved Heating of Powder Metal Impacts," *IEEE Trans. Ind. Applications IA-13* (4):330 (1979).
42. U.S. Department of Commerce, Bureau of the Census. *1977 Census of Manufacturers: Preliminary Reports,* Washington, DC (1979).
43. Cline, H. E., and T. R. Anthony. "Heat Treating and Melting Material With a Scanning Laser or Electron Beam," *J. Appl. Phys.* 48(9):3895-3900 (1977).
44. Krouse, J. K. "Laser Metalworking," *Machine Design* (November 9, 1978).
45. Spalding, I. J. "High Power Lasers for Processing of Materials—A Comparison of Available Systems," *Optics Laser Technol.* 29-32 (February 1978).
46. Laser Applications, Inc. Personal Communication (December 18, 1979).
47. Dale, B. W. "Looking Into Engines With Lasers," *Ind. R&D* 121-124 (November, 1979).
48. Hutzel, J. "Laser System Improves Testing Experimental Alloys," *Ind. R&D* 22(2):94-95 (1980).
49. Langevin, J. M. "Short-Pulse Laser Drills Holes Without Taper," *Ind. Wk.* 26 (March 20, 1978).
50. Price, P. "Faster, Stronger, Machining Center," *Prod. Eng.* 29 (June 1980).
51. Donyina, K., J. D. Lavers and R. S. Segsworth. "Plasma Processing of Ferromanganese Slabs," paper presented at the Electric Furnace Conference, Detroit, MI, December 4-7, 1979.
52. Miller, J. E., and J. A. Wineman. "Laser Hardening at Saginaw Steering Gear," *Metal Prog.* 38-43 (May 1977).
53. Siemens, F. "Laser Beams as Tools for Heat Treating," paper presented at the SME Heat Treatment Conference, Los Angeles, CA, June 10-12, 1980.
54. Robinson, A. L. "Laser Annealing: Processing Semiconductors Without a Furnace," *Science* 201:333 (July 28, 1978).
55. Fairchild, J. M., and G. H. Schwuttke. *Solid State Electronics* 11:1175 (1968).
56. Hess, L. D., and G. Yaron. "Laser Annealing," *Ind. R&D* 141-152 (November 1979).
57. Hutzel, J. "Laser Used to Fabricate 'Supersaturated' Alloys," *Ind. R&D* 22(5):60 (1980).
58. Dreger, D. R. "Pinpoint Hardening by Electron Beam," *Machine Design* 89-93 (October 26, 1978).
59. Jenkins, J. E. "Electron Beam Surface Hardening," *Tooling Prod.* 76-77 (December 1978).
60. Gilbert, C. L. "Computerized Control of Electron Beam for Precision Surface Hardening," *Ind. Heating* 45(1):16-18 (1978).
61. Bianchi, L. M. "Electron Beam PVD Corrosion Resistant Coatings for Extended Life of Gas Turbine Parts," *Ind. Heating* 47(6):24-27 (1980).
62. Hegland, D. E. "Tomorrow's Technology Builds Jet Engines Today," *Prod. Eng.* 27(7):30-35 (1980).
63. Marcellin, W. "Application for Electron Beam Welding," *Society for Manufacturing Engineering Conference on Heat Treatment,* Los Angeles, CA, June 10-12, 1980.

64. Kelly, M. "Thermal Processing of Steel Strip Grades at New Computerized Cold Rolling Mill Related to Specific Applications," *Ind. Heating* 46(9):20-25 (1979).
65. Graham, T. C. "Jones and Laughlin Electric Steelmaking," paper presented at the Electric Furnace Conference, Detroit, MI, December 4-7, 1979.
66. Greggerman, I. "National View of Productivity," paper presented at the Electric Furnace Conference, Detroit, MI, December 4-7, 1979.
67. Hillegas, A. L. "Steel Industry View of Productivity," paper presented at the Electric Furnace Conference, Detroit, MI, December 4-7, 1979.
68. *United Nationsl Statistical Yearbook for 1977* (1978).
69. Lubbeck, W. "The Effect of Electric Power and the Consumption of Energy in High Performance Electric Arc Furnaces," paper presented at the 14th National Industrial Electric Heat Treatment Conference, Louisville, KY, March 3-5, 1980.
70. Galbreath, J. G. "Electric Arc Furnaces in Mini-Mills in the United States," paper presented at the 14th National Industrial Electric Heat Treatment Conference, Louisville, KY, March 3-5, 1980.
71. Barcza, N., et al. "Excavation of a 75 MVA High-Carbon Ferromanganese Electric Smelting Furnace," paper presented at the Electric Furnace Conference, Detroit, MI, December 4-7, 1979.
72. Breton, E. "Maze of Ferrosilicon Smelting," paper presented at the Electric Furnace Conference, Detroit, MI, December 4-7, 1979.
73. Guile, A. E. "Arc-electrode Phenomena." *IEE Rev. (Great Britain)* 118 9R:1131-1154 (September 1971).
74. Finkelnburg, W. "The High Current Carbon Arc and Its Mechanism," *J. Appl. Phys.* 20:468-474 (May 1979).
75. Rennie, M. B. "Production Models for High-Carbon Ferrochromium," paper presented at the Electric Furnace Conference, Detroit, MI, December 4-7, 1979.
76. Olsen, L., and R. Innvaer. "Mathematical Models for Soderberg Electrodes," paper presented at the Electric Furnace Conference, Detroit, MI, December 4-7, 1979.
77. Stewart, A. B. "Submerged Arc Furnace Electric Circuit Analysis," paper presented at the Electric Furnace Conference, Detroit, MI, December 4-7, 1979.
78. Baker, G. M. "Electrical Considerations in Electric-Arc Furnace Productivity," paper presented at the Electric Furnace Conference, Detroit, MI, December 4-7, 1979.
79. Fairchild, W. T. "Production of Silvery Pig Iron in Covered Submerged Arc Furnaces, paper presented at the Electric Furnace Conference, Detroit, MI, December 4-7, 1979.
80. Nafziger, R. H. "Ilmenite Reduction by a Carbon Injection Technique," paper presented at the Electric Furnace Conference, Detroit, MI, December 4-7, 1979.
81. Cundiff, C. L. "Update of the Melting and Continuous Casting Operations at Armco," paper presented at the Electric Furnace Conference, Detroit, MI, December 4-7, 1979.
82. Lectromelt Corporation. "Vacuum Arc Melting Furnaces," *LECTROMELT Catalogue* #13 (1979).
83. Ray, C. T., and W. L. Wilburn. "Operation of a Two-Furnace Ferrosilicon Plant Under Process Computer Control," paper presented at the Electric Furnace Conference, Detroit, MI, December 4-7, 1979.

84. Wilbern, W. L. "Computer Control of Submerged-Arc Ferroalloy Furnace Operations," *Electric Furnace Proc.* (1974).
85. Blevins, S. "Productivity of the Steelton Electric Furnace," paper presented at the Electric Furnace Conference, Detroit, MI, December 4-7, 1979.
86. Kroncke, G., et al. "Productivity of the Steelton Electric Furnace," paper presented at the Electric Furnace Conference, Detroit, MI, December 4-7, 1979.
87. Alcock, C. B., and R. S. Segsworth. "Extended Arc Furnace and Process for Melting Particulate Charge Therein," U.S. Patent #4006284 (1977).
88. Grube, W. L. "High Rate Plasma Carburizing With Propane," *J. Heat Treating* 1(1):95-97 (1979).
89. Grube, W. L., and J. G. Gay. "High-Rate Carburizing in a Glow-Discharge Methane Plasma." *Metallurgical Trans. A* 9A:1421-1429 (October 1978).
90. Grube, W. L. "Direct-Current Glow-Discharge Furnace for High-Rate Carburizing," *J. Vacuum Sci. Technol.* 16(2):335-338 (1979).
91. Grube, W. L. "High Rate Plasma Carburizing With Propane," *J. Heat Treating* 1(1):95-97 (1979).
92. Grube, W. L. Personal communication (1980).
93. Duff, D. B. "Plasma Smelting Is Key to Future of Platinum Project," *Eng. Mining J.* 38-43 (September 1979).
94. Davidson, H. "Tungsten Mesh Heating Elements for High Temperature Applications," *Wire J.* 10(10):84-87 (1977).
95. Hutzel, J. "New 'Porcupine' Element Makes Heat Guns Safer and More Efficient," 46(12):22 (1979).
96. Slater, K. A. "Fundamental and Typical Applications of Electric Immersion Heaters," paper presented at the 14th National Industrial Electric Heat Treatment Conference, Louisville, KY, March 3-5, 1980.
97. Hutzel, J. "Plasma Spray Coating Increases Service Life of New Immersion Heaters," *Ind. Heating* 46(12):18 (1979).
98. Vanderkaay, D. H. "Conversion of Furnaces Using Ceramic Fiber Modules and New Electric Heating Systems," *Ind. Heating* 47(1):34 (1980).
99. Bauer, M. K. "Ceramic Fiber Heater . . . A New Concept in Building Lightweight Electric Furnaces," paper presented at the 14th National Industrial Electric Heat Treatment Conference, Louisville, KY, March 3-5, 1980.
100. Romano, P. "Autoclave at Aircraft Plant Switched from Oil/Gas to Electric Resistance Heat," *Ind. Heating* 45(7):18-20 (1978).
101. Ward, J. "Recent Developments in Heat Treatment of Ductile Iron," paper presented at the 5th ASM Heat Treatment Workshop, Detroit, MI May 6-8, 1980.
102. Otto, G. "Sintering of Stainless Steels," paper presented at the 5th ASM Heat Treatment Workshop, Detroit, MI, May 6-8, 1980.
103. Brewer, J. Personal Communication (May 19, 1980).
104. Hutzel, J. "Annealing of Cold Drawn Steel in New Electrically Heated Cover Type Furnaces," *Ind. Heating* 46(9):35 (1979).
105. Hutzel, J. "Numerically Controlled Gear Tooth Induction Hardening Machine Heat Treats with Minimum Distortion," *Ind. Heating* 47(3):24 (1980).
106. Leonard, L. "Metalcasting Shapes Up," *Prod. Eng.* 25(6):51-54 (1978).
107. Widmer, R. "Hot Isostatic Pressing," paper presented at the Advances in Carbide Tooling Seminar, Detroit, MI, May 5, 1977.

108. Widmer, R. "Densification of Castings by Hot Isostatic Pressing," paper presented at the 26th Annual Meeting Investment Casting Institute, October, 1978.
109. Price, P. "Hot Isostatic Pressing–New Technology," paper presented at the 5th ASM Heat Treatment Workshop, Detroit, MI, May 6-8, 1980.
110. Price, P. "Hot Isostatic Pressing–A New Heat Treating Technology with Tremendous Potential," *Ind. Heating* 46(6):8-10 (1979).
111. Hutzel, J. "Pipe Tracing–Fundamentals and Proper Applications," paper presented at the 14th National Industrial Electric Heat Treatment Conference, Louisville, KY, March 3-5, 1980.
112. Koester, G. L. "Electric Resistance Process Heating," *Plant Eng.* 32(3): 69-72 (1978).
113. Durant, J. H. "Heat Transfer in Vacuum Furnaces–Its Relationship to Productivity and Energy," paper presented at the ASM Heat Treating Conference, Detroit, MI, May 24-26, 1977.
114. Warner, J. C. "Vacuum Brazing of Aluminum and Copper," paper presented at the 5th ASM Heat Treatment Workshop, Detroit, MI, May 6-8, 1980.
115. Warner, J. C., and W. C. Weltman. "The Fluxless Brazing of Aluminum Radiators," paper presented at the Society of Automatic Engineers, Paper #780299, February 27-March 3, 1978.
116. Szekely, J. "Energy Conservation Via Carburizing," *Tooling Prod.* 80-81 (December 1977).
117. Hutzel, J. "Energy Conservation by Means of Vacuum Furnace Carburizing," *Ind. Heating* 47(7):13-14 (1980).
118. Masters, C. F. "Electric Vacuum Furnaces for Heat Treating Steel," *Ind. Heating* 45(9):23-28 (1978).
119. Ross, J. A., and R. D. Lobell. "Application of Vacuum Heat Treating at Commercial Heat Treating Plant," *Ind. Heating* 47(1):12-16 (1980).
120. Hutzel, J. "Use of Ceramic Fiber in Vacuum Heat Treating-Furnaces," *Ind. Heating* 47(5):44 (1980).
121. Hopkins, R. K. "Manufacture of Alloy Ingot," U.S. Patent #2191479 (February 1940).
122. Mitchell, A. "Electroslag Casting–Its Future in America," *Modern Casting* 86-88 (November 1978).
123. Schlatter, R. "Direct Current Electroslag Remelting of Specialty Alloy Steels," *Ind. Heating* 45(10):19 (1978).
124. Philip, T. V. *Metals Technol.* 2(12):554-564 (1975).
125. Hill, R. M., and F. Wilson. "Direct Electric Resistance Heating in the Forge Shop," paper presented at the 14th National Industrial Heat Treatment Conference, Louisville, KY, March 3-5, 1980.
126. Nankin, M. "Internal Electrothermic Treatment Steels and Alloys," paper presented at the 5th ASM Heat Treatment Workshop, Detroit, MI, May 6-8, 1980.
127. Baldwin, S. Personal Communication (July 31, 1980).
128. Lord, N. W., and B. Borko. "Selection of Steam or Electric Energy in U.S. Industry, The MITRE Corporation, Technical Report Number MTR-7209, Series 15 (1977).
129. Marley, M. "Should You Farm Out Energy Intensive Work," *Iron Age* (December 3, 1979).
130. Guile, A. E. "Outlook '80," *Heat Treating* (January 1980).
131. Hutzel, J. "Industrial Heating Equipment Sales Expected to Increase 28% Over 1979," *Ind. Heating* 47(2):6 (1980).

INDEX

157